GRADE 6 MATH REVIEW & PRACTICE BOOK

Exceed in Math

BY
Dr. Mulugeta Markos

Copyright © 2020 by Mulugeta Markos

ISBN: 9798669978501

All rights reserved. No part of this book may be reproduced or transmitted in any form or by any means, electronic or mechanical, including photocopying, recording, or by any information storage, without permission in writing from the owner.

To order additional copies of this book, contact

getthemathtutoring@gmail.com

or

www.amazon.com and use ISBN: 9798669978501

About this book

This Review Notes and Practice Workbook is intended to prepare students to exceed in the classroom and in the next grade levels. The book is systematically organized to include all types of mathematical problems in the sixth grade level. Each chapter is designed to comprise: review notes, basic practice, challenge problems, application word problems, and proficiency test. At the end you will find five standardized exam practices. The workbook has about 3,000 practice problems with a complete answer key.

The book is developed by an expert in the field, Ph.D. in mathematics, with more than twenty years of teaching experience. The author prepared the book considering students need to be equipped with the necessary skills they need to succeed in the classroom and beyond. All problems in this practice book are classroom-tested and are prepared to fulfill state and national content standards. It is clear that when an expert in the field, involved in teaching for long period of time, prepares the book, students will benefit a lot from the author's immense expertise. The book has eleven chapters. At the beginning of each chapter, you will find short notes for review purposes, then series of problems divided by sections and subsections covering all topics in the six grade.

The book is extremely helpful for parents seeking additional work for their children. Parents can easily assign extra practice assignments from this book that matches students' classroom lessons. Answer keys are provided at the end of the book which allow parents to check students' work. In order to excel in six grade math, each student is encouraged to make a 90% or above in all assigned problems and chapter tests.

TABLE OF CONTENTS

CHAPTER 1   Page
OPERATIONS WITH DECIMALS

1.0 Review Notes on Decimals ………………………………………………...1

1.1. Adding & Subtracting Decimals …………………………… …. 2

1.2. Estimating Sums and Differences ……………………………. 3

1.3. Multiplying Decimals…………………………………………. 4

1.4. Dividing Multi-Digit Numbers………………………………... 5

1.5 Dividing and Multiplying by a power of 10 ………………….. 6

1.6. Estimating Products and Quotients……………………………7

1.7. Dividing Decimal …………………………………………….7

1.8. Comparing and Ordering Decimals …………………………..8

1.9. Word Problems on Decimals ………………………………… 9

1.10. Test …………………………………………………………..13

CHAPTER 2

FRACTION

2.0 Review Notes on Fractions …………………………………..............18

2.1 Add and Subtract Fractions ….…………………… ………………19

2.2. Estimate product of Fractions ………………………………………...24

2.3. Proper and Improper Fractions ………………………………………. 24

2.4. Mixed to Improper Fraction ……………………………………………..25

2.5. Improper Fraction to Mixed number ……………………………………27

2.6. Add & Subtract Mixed numbers …………………………………………28

2.7. Multiply Fractions ……………………………………………… …..30

2.8. Multiplying Mixed Numbers …………………………………………...34

2.9. Dimensional analysis …………………………………………………35

2.10. Divide Fractions ………………………………………………………..37

2.11 Dividing Mixed Numbers ………………………………………………….40

2.12. Prime Factorization ………………………………………………...........42

2.13 Greatest Common Factor ……………………………………………… 44

2.14. Least Common Multiple …………………………………………….… 44

2.15 Simplifying Fractions ……………………………………………….............45

2.16. Reducing Fractions to Lowest Form …………………………….……….46

2.17 Mixed Numbers with Mixed Operations ………………………….…… 49

2.18 Ordering Fractions …………………………………………………….50

2.19 Word problems Involving Fractions …………………………………….54

2.20 Test …………………………………………………………………….64

CHAPTER 3
RATIOS, PERCENT & PROPORTIONS

3.0 Review Notes on Ratios, Percent and Proportions ………………………69

3.1 Ratios …………………………………………………………………..70

3.2. Rates and Unit rates ……………………………………………………..74

3.3. Equivalent Ratios and Proportions ……………………… …………….74

3.4 Ratio tables and Graphs………………………………… …………….77

3.5. Percent to Fraction ……………………………………………………...79

3.6. Percent to Decimal ……………………………………………………..80

3.7. Relate Fractions, Decimals, and Percent ……………………… ……….81

3.8. Finding Percent of a Number …………………………………………...84

3.9. Comparing and ordering Percent, Fractions, and Decimals……………..85

3.10. Estimating with percent ………………………………………………86

3.11. Percent Word Problems ……………………………………………….87

3.12. Ratio & Proportion Word Problem …………………………………...92

3.13. Test …………………………………………………………………….97

CHAPTER 4
EXPRESSIONS & EQUATIONS

4.0. Review Notes on Expressions & Equations..103

4.1. Exponent and product..104

4.2. Order of Operation...106

4.3. Variables, Expressions and Equations..107

4.4. Evaluating Expressions 109

4.5. Writing Algebraic Expression...110

4.6. Properties 111

4.7. Equivalent Expressions...112

4.8. Factoring Expressions..116

4.9. Simplifying Expressions with One and Two Variables.....................114

4.10. Solving Equations: Addition Principle 119

4.11. Solving Equations: Multiplication. Principle 121

4.12. Solving Multistep Equations 124

4.13. Formulas 126

4.14. Expression Word Problems 127

4.15. Test 129

CHAPTER 5

INEQUALITIES AND FUNCTIONS

5.0 Review Notes on Functions and Inequalities 134

5.1 Definition of function 134

5.2 Function tables and rules..135

5.3 Arithmetic and Geometric Sequences..138

5.4. Inequalities..139

5.5. Writing Inequalities..141

5.6. Graphing Inequalities...141

5.7. Solving Inequalities..143

5.8. Word Problem Involving Inequalities and Functions........................145

5.9. Test...146

CHAPTER 6
AREA

6.0 Review Notes on Area of Polygons……………………………………………151

6.1 Properties of Polygons ……………………………………………… ……….152

6.2. Area of Parallelograms ……………………………………………………...153

6.3. Area of Triangles …………………………………………………………...155

6.4. Area of Trapezoids …………………………………………………………156

6.5. Area of Circles …………………………………………………………...157

6.6. Area of Composite Figures ………………………………………………159

6.7. Word Problems on Perimeter and Area ……………………………………161

6.8. Test …………………………………………………………………………163

CHAPTER 7
SURFACE AREA AND VOLUME

7.0. Review Notes on Surface Area and Volume……………………………….167

7.1. Surface Area………………………………………………………………..168

7.2. Three Dimensional Figures………………………………………………...170

7.3. Volume……………………………………………………………………..171

7.4. More on Surface Area and Volume………………………………………...173

7.5. Word problems on Surface Area and Volume……………………………..174

7.6. Test…………………………………………………………………………176

CHAPTER 8
STATISTICS AND PROBABILITY

8.0 Review Notes on Statistics ………………………………………………….181

8.1 Statistical Questions…………………………………………………………182

8.2. Measures of Center ………………………………………………………… 183

8.3. Measures of Variation………………………………………………………192

8. 4. Outliers…………………………………………………………………….196

8.5. Box and whisker Plot………………………………………………………198

8.6. Line plot …………………………………………………………………....199

8.7. Bar Graph……………………………………………………………….200

8.8. Line Graphs …………………………………………………………….203

8.9. Frequency Distribution and Histogram……………………………………205

8.10. Circle Graphs ……………………………………………………….. .207

8.11. Compared Data Set ……………………………………………………..210

8.12. Basic Probability Properties……………………………………………..213

8.13. Probability Experiment & Outcomes ……………………………………...213

8.14. Probability as a Percent ………………………………………………...215

8.15. Probability as a Fraction ……………………………………………….215

8.16. Word Problems on Statistics and Probability ………………………… 217

8.17. Test ………………………………………………………………… 202

CHAPTER 9
INTEGERS

9.0 Review Notes on Integers …………………………………………….228

9.1. Integers ……………………………………………………………….229

9.2 Opposite Integers……………………………………………………...230

9.3. Comparing Integers …………………………………………………….231

9.4. Add & Subtract Integers ……………………………………………….234

9.5. Coordinate Plane ……………………………………………………….234

9.6 Quadrants and Axis ……………………………………………………...235

9.7. Graph Order Pairs ……………………………………………………....236

9.8. Function Rule & Graph ………………………………………………….238

9.9. Map and Geometry on Coordinate Planes ……………………………….239

9.10. Word Problems on Integers …………………………………………….239

9.11. Test ……………………………………………………………………243

CHAPTER 10
RATIONAL NUMBERS

10.0 Review Notes on Rational Numbers 248

10.1 Terminating & Repeating decimal 248

10.2. Compare and Order Rational Numbers 249

10.3. Absolute Value ……………………………………………………251

10.4 Distance on a Number Line and Coordinate Plane............... 253

10.5 Polygons on Coordinate Planes……………………………………..254

10.6. Add & Subtract Rational Numbers………………………….. …..257

10.7. Multiply & Division Rational Numbers ……………………….257

10.8. Word Problems on Rational Numbers …………………………...258

10.9. Test ……………………………………………………………….260

CHAPTER 11

PRACTICE EXAMS ………………………………………………….266

ANSWERS ……………………………………………………………316

CHAPTER 1

OPERATIONS WITH DECIMALS

1.0 Review Notes on Decimals

- A decimal number has three parts: the whole number part, the point and the decimal part. The first place after the decimal point is obtained by dividing the number by 10; it is called the tenths place. The second place after the decimal point is obtained by dividing the number by 100; it is called the hundredths place. The third place after the decimal point is obtained by dividing the number by 1000; it is called the thousandths place, and so on.

- To add or subtract decimals line up the decimal points and add or subtract the digits in the same positions. Put zero to hold any missing place values.

- Multiplying by powers of 10 is moving the decimal point to the right or left by the number of zeros in the power of 10. When multiplying by numbers like 1/10, 1/100, we move the decimal point to the left. When multiplying by numbers like 10, 100, 1000, we move the decimal point to the right.

- When multiplying decimals count the number of digits after the decimal point in each number, add them, then multiply the numbers as you multiply whole numbers. After multiplying put the decimal point in the answer - it will have as many decimal places as the two original numbers combined.

- To write mixed number as a decimal, convert the fraction part of a mixed number to a decimal and add to the whole number.

- To round a decimal to a specific place value, look at the digit next to that specific place value to the right. If it's 4 or less, just remove all the digits to the right. If it's 5 or greater, add 1 to that certain digit place, and then remove all the digits to the right.

- To estimate sums, differences, and products of decimals:

    Step 1: Round each decimal to the required digit

    Step 2: Perform the required operation

    Step 3: Round the answer to required digit.

- To estimate quotient of decimals use rounding the numerator or denominator so that the result of the quotient can be calculated easily.

- When comparing decimals, start in the tenths place. The decimal with the biggest tenth value is greater. If the tenth places are the same, move to the hundredths place and compare these values. If the values are still the same keep moving to the right until you find one that is greater or until you find that they are equal.

- To divide a decimal by a whole number, divide as a whole number and place the decimal in the quotient right above its place in the dividend. To divide by a decimal denominator first change the divisor to the whole number by multiplying both the divisor and the dividend by the same power of 10. Then use the method of dividing a decimal by a whole number.

1.1 Adding and Subtracting Decimals

1.1.1. Add

a. $57.2 + 43.5 =$ _____

b. $6,784.11 + 4,266.89 =$ _____

c. $8.886 + 7.05 =$ _____

d. $3.874 + 2.648 =$ _____

e. $2387.4 + 2648 =$ _____

f. $3.127 + 0.598 =$ _____

g. $134.21 + 396.36 =$ _____

h. $38.7478 + 0.2648 =$ _____

i. $0.989807 + 1.385460 =$ _____

j. $1,346.559 + 230.426 =$ _____

k. $999,999.9 + 100,000.1 =$ _____

l. $133,460.89 + 932,146.7 =$ _____

1.1.2. Subtract

a. $1,027.2 - 593.5 =$ _____

b. $5,784.11 - 4,266.89 =$ _____

c. $78.886 - 7.05 =$ _____

d. $6.874 - 2.648 =$ _____

e. 7,387.4 – 2,648.3 = _____

f. 3.127 – 2.598 = _____

g. 734.21 – 396.36 = _____

h. 38.7478 – 0.2648 = _____

i. 7.989807 – 1.385460 = _____

j. 1,346.569 - 230.426 = _____

k. 999,999.9 – 100,000.1 = _____

l. 933,460.89 - 932,146 = _____

1.2. Estimating Sums and Differences

1.2.1. Round each decimal to the nearest whole number.

a. 2.46 = _____

b. 12.61 = _____

c. 4.99 = _____

d. 3.01 = _____

e. 4.53 = _____

f. 213.17 = _____

g. 22 = _____

1.2.2. Estimate each sum or difference to the nearest whole number.

a. 3.4 + 7.84 = _____

b. 21.88 + 451.1 = _____

c. 200.2 - 11.78 = _____

d. 12.9 + 14.8 = _____

e. 26.006 - 1.78 = _____

f. 88.6 - 12.76 = _____

g. 9.99 + 19.09 = _____

1.2.3. Estimate each sum or difference to the nearest tenths

a. 12.08 + 4.1= _____

b. 4.56 + 2.34 = _____

c. 123.67 + 54.78 = _____

d. 77. 02 – 24.10 = _____

e. 88.88 – 22.22 = _____

f. 105.0 – 26. 02 = _____

1.3. Multiplying Decimals

1.3.1 Multiply each decimal by a whole number

a. 4 × 2.5 = _____

b. 7 × 4.5 = _____

c. 9 × 0.23 = _____

d. 3 × 4.44 = _____

e. 7 × 9.7 = _____

f. 77 × 0.77 = _____

g. 10 × 3.5 = _____

h. 6 × 33.5 = _____

i. 80 × 0.90 = _____

j. 88 × 6.6 = _____

k. 45 × 5.4 = _____

l. 99 × 9.9 = _____

1.3.2 Multiply decimal by decimal

1. 1.2 × 3.1 = _____

2. 3.2 × 0.03 = _____

3. $4.6 \times 2.8 =$ _____

4. $3.01 \times 6.6 =$ _____

5. $4.21 \times 5.7 =$ _____

6. $2.61 \times 0.11 =$ _____

7. $12.6 \times 23.1 =$ _____

8. $12.23 \times 35.67 =$ _____

9. $23.68 \times 80.14 =$ _____

10. $123.3 \times 0.1 =$ _____

11. $12.3 \times 0.01 =$ _____

12. $32.1 \times 0.001 =$ _____

13. $26.45 \times 0.0001 =$ _____

14. $0.1 \times 0.01 =$ _____

15. $0.01 \times 0.001 =$ _____

16. $2.01 \times 0.0001 =$ _____

17. $0.11 \times 0.111 =$ _____

18. $3.01 \times 0.232 =$ _____

19. $8.08 \times 1.12 =$ _____

20. $32.01 \times 0.324 =$ _____

1.4. Dividing Multi-Digit Numbers

a. $84 \div 21 =$ _____

b. $600 \div 15 =$ _____

c. $720 \div 80 =$ _____

d. $2490 \div 30 =$ _____

e. $315 \div 15 =$ _____

f. $705 \div 15 =$ _____

g. $364 \div 26 =$ _____

h. $696 \div 12 =$ _____

i. $783 \div 27 =$ _____

j. $756 \div 42 =$ _____

k. $874 \div 38 =$ _____

l. $884 \div 26 =$ _____

1.5. Dividing and multiplying Decimals by a power of 10

Divide or Multiply

a. $53.75 \times 10 =$ _____

b. $2.05 \times 100 =$ _____

c. $2.123 \times 10,000 =$ _____

d. $0.9876 \times 1,000 =$ _____

e. $34.24 \times 0.1 =$ _____

f. $1.34 \times 0.01 =$ _____

g. $640 \times 0.0001 =$ _____

h. $0.31 \times 0.001 =$ _____

i. $53.75 \div 10 =$ _____

j. $2.05 \div 100 =$ _____

k. $212.3 \div 10,000 =$ _____

l. $0.9876 \div 1,000 =$ _____

m. $34.24 \div 0.1 =$ _____

n. $1.34 \div 0.01 =$ _____

o. $0.640 \div 0.0001 =$ _____

p. $0.31 \div 0.001 =$ _____

q. $4005 \div 1,000 =$ _____

r. $0.00001 \times 100,000 =$ _____

s. $10,000 \div 10,000 =$ _____

t. $0 \div 10,000 =$ _____

1.6. Estimating Products and Quotients

Estimate each product or quotient by rounding to the nearest whole number.

a. $3.8 \times 4 =$ _____

b. $4.2 \times 10.1 =$ _____

c. $9.9 \times 3 =$ _____

d. $4.01 \times 10 =$ _____

e. $5.75 \times 6 =$ _____

f. $6.2 \times 7.8 =$ _____

g. $8.1 \times 8.9 =$ _____

h. $3.95 \times 2.8 =$ _____

i. $9.99 \times 0.001 =$ _____

j. $11.9 \div 3 =$ _____

k. $24 \div 2.9 =$ _____

l. $9.75 \div 5 =$ _____

m. $14.87 \div 3 =$ _____

n. $8.9 \div 9 =$ _____

o. $11.98 \div 2.98 =$ _____

p. $7.8 \div 2.01 =$ _____

q. $89.9 \div 29.9 =$ _____

r. $0.01 \div 21 =$ _____

1.7. Dividing Decimal

1.7.1. Divide each decimal by a whole number

1. $12.4 \div 4 =$ _____

2. $8.2 \div 2 =$ _____

3. $13.6 \div 4 =$ _____

4. $22.5 \div 9 =$ _____

5. 10.5 ÷ 21 = _____

6. 19.6 ÷ 4 = _____

7. 13.84 ÷ 2 = _____

8. 1.8 ÷ 2 = _____

9. 6.16 ÷ 11 = _____

10. 14.81 ÷ 1 = _____

11. 19.2 ÷ 8 = _____

12. 17.15 ÷ 7 = _____

13. 30.4 ÷ 8 = _____

14. 20.1 ÷ 10 = _____

15. 1700 ÷ 4.25 = _____

16. 100 ÷ 0.5 = _____

17. 300 ÷ 0.3 = _____

18. 10 ÷ 0.1 = _____

19. 7.5 ÷ 3 = _____

20. 7.5 ÷ 30 = _____

1.7.2 Divide each by the given decimal

1. 2.88 ÷ 0.4 = _____

2. 193.2 ÷ 0.84 = _____

3. 16.24 ÷ 0.07 = _____

4. 1.36 ÷ 6.8 = _____

5. 4.14 ÷ 1.8 = _____

6. 2.16 ÷ 5.4 = _____

7. 8.4 ÷ 0.01 = _____

8. 0.484 ÷ 0.2 = _____

9. 100.8 ÷ 4.8 = _____

10. 678.4 ÷ 5.0 = _____

11. 54.4 ÷ 3.2 = _____

12. 51.84 ÷ 7.2 = _____

13. 0.2 ÷ 0.08 = _____

14. 1.488 ÷ 0.24 = _____

15. 1.488 ÷ 6.2 = _____

16. 31 ÷ 2.5 = _____

17. 31 ÷ 12.4 = _____

18. 21.125 ÷ 3.25 = _____

19. 21.125 ÷ 6.5 = _____

20. 0.25 ÷ 0.01 = _____

1.8. Comparing and ordering decimals

1.8.1. Compare the decimals. (Use <, >, or =)

1. 0.101 _____ 0.11

2. 23.51 _____ 21.99

3. 2.90 _____ 3.078

4. 0.678 _____ 0.659

5. 123.0082 _____ 123.0121

6. 1.2 + 3.2 _____ 2.2 × 2.5

7. 4.01 − 3.01 _____ 2.01 ÷ 2

8. 20 ÷ 50 _____ 0.4

9. 2 × 0.001 _____ 10.01

1.8.2. Order the decimals from least to greatest

1. 0.3, 0.08, 1.2, 3.45 = _____

2. 0.4325, 4.321, 0.04992, 2.0432 = _____

3. 55.5, 5.05, 50.05, 55.505 = _____

4. 21.11, 11.11, 21.011, 21.01 = _____

5. 30.0003, 30.0016, 29.9993, 30.00009 = _____

6. 0.4312, 0.4213, 0.4321, 0.4132 = _____

7. 0.8, 0.04213, 0.4312, 0.0013 = _____

1.9. Word Problems Involving Decimals

1. On average a bicyclist rides 45.5 miles per hour. About how many miles does the bicyclist ride in 2.45 hours? = _____

2. A T-shirt costs $19.99. How much change will you receive if you pay a $100 bill? = _____

3. A Boeing – 747 can fly 614.5 miles per hour. Calculate the distance that this Boeing can travel in 10.5 hours? Assume the same speed. = _____

4. The average height of a tall tree is 22.4ft. The average height of a short tree is 16.8ft. Estimate the difference of their heights. = _____

5. The average height of a tall tree is 22.4ft. The average height of a short tree is 16.8ft. Find the sum of their heights. = _____

6. A whole food store sells 10 pound of fish meat for $99.99. How much would a pound of fish cost? = _____

7. Amy makes $7.75 per hour at Chick-fil-A. If she works 7 hours a day, how much money will she make in five days? = _____

8. At a book fair, a 5th grade math book costs $9.95. A teacher bought 28 math books for 5th graders. How much money did the teacher spend? = _____

9. Tom weighs 120.43 pound. His son, Bob, weighs one-fourth as much as his dad. How much does Bob weighs? = _____

10. You want to stack 25 bricks. Each brick is 0.78 ft. How many ft. tall would your bricks be if the bricks are stack one on the top of the other? = _____

11. Joe eats 3.32 pound of apple in 8 days. How much is his daily consumption if Joe eats equal amount every day? = _____

12. The distance from Britany's house to her school is 13.56 miles. How many miles does Britany travel in a week assuming she goes five days to school in a week? = _____

13. XYZ Elementary School needs 24 school buses for a daily transportation. Each school bus uses 14.19 gallons of gasoline a day. How much fuel is needed for a day? = _____

14. Multiply 2.34 by 4.12 then divide the answer by three. Find the final result to the nearest tenth. = _____

15. Think of a number, add 2.38 to the number and subtract 0.38 the result 3. What is the number? = _____

16. Bill scored 84.38 out of 100 in math. The teacher put one wrong problem in the test, as a result he decided to add 4.18 point for each student. Calculate Bill's new score. = _____

17. The weight of a new born baby boy is 7.62 pounds, after two months he doubled his weight. After that, his weight increased by 0.78 pounds for 6 consecutive months. What is the boy's weight after 8 months? = _____

18. The side length of a square is 12.34 inches. Find the perimeter? = _____

19. The area of a rectangular backyard is 33.4772 square meters and its length is 13.18 meters. Find the width of the backyard = _____

20. The weight of a sack of sand is 48 kilograms. If you want to distribute it into buckets caring 1.98 kilograms each. Approximately how many buckets do you need to distribute the sack of sand? = _____

21. A soccer team of 20 players need new pairs of pants, shirts, and pairs of shoes. The cost of each pair of pants, shirt, and pair of shoes are $13.42, $18.5 and $26.25, respectively. How much money is needed to cover the cost for the 20 players? = _____

22. It costs $6.85 to ship a 3.1pound book. About how many such books would you ship for $210? = _____

23. Mark can run 5.65 miles per hour. At this rate how far could Mark run in 90 minutes?

    = _____

24. Ronaldo runs 6.5 miles per hour. Messi runs 24 miles in 2.5 hours. Who is faster?

= _____

25. In question 24, if both run for 2.5 hours. How many miles would both run together?

= _____

1.10. Chapter 1 Test

1. Comparing: 0.094 _____ 1.0 – 0.906

   a. >
   b. <
   c. =

2. Mark weighs 213.28 pounds in the morning and 218.32 pounds at night. How many more pounds did Mark weigh at night than in the morning?

   a. 5.04
   b. 5.40
   c. 321.6
   d. 208.28

3. What is 14.11+ 11.26 - 8.80?

   a. 6.75
   b. 16.57
   c. 15.36
   d. 9.89
   e. 10.11

4. Divide 0.45 ÷ 1,000

    a. 45
    b. 0.00045
    c. 0.045
    d. 4500

5. Multiply 835.6 × 0.01

    a. 8346
    b. 835.6
    c. 83.56
    d. 8.356

6. Multiply 0.34 × 2.32

    a. 0.7888
    b. 78.8
    c. 7.888
    d. 7888

7. Divide 89.4 ÷ 50

    a. 7.88
    b. 17.88
    c. 1.78
    d. 1.788

8. The area of a square yard is 475.24 square meters and its length is 21.8 meters. Find the width of the yard?

   a. 21 meters
   b. 21.4 meters
   c. 21.8 meters
   d. 21.6 meters

9. Select a decimal equivalent to three – tenths

   a. 5.0
   b. 0.8
   c. 0.3
   d. 0.5

10. What is the estimated sum to the nearest whole number for 4.61 + 1.99?

    a. 7
    b. 6
    c. 5
    d. 5.6

11. Sixty-one and twelve thousandths in standard form is:

    a. 61.012
    b. 61.12
    c. 61.0012
    d. 61.22

12. $100 + 4 \times 8 + 3 \times 0.3 - 5 \times 0.001$ in standard form is:

   a. 132.895
   b. 132.905
   c. 132.9005
   d. 132.95

13. $3.6 \div 500 \times 1,000$

   a. 72
   b. 7.2
   c. 0.072
   d. 0.0072

14. Divide 81.45 by 6 and round the quotient to the nearest tenth

   a. 13.6
   b. 13.5
   c. 13.7
   d. 13.8

15. $4.2 \div 2.1 - 1 + 1$ is

   a. 2
   b. 1
   c. 0
   d. 2.1

16. A page in a book has 450 words on it. Assume each page has equal number words. If the book has 288 pages. How many words are on the book?

a. 219, 600

b. 12, 960

c. 129, 600

d. 129,060

17. John paid $ 14.7 for 7.5 pounds of oranges. What is the cost for two pounds of oranges?

   a. $3.92
   b. $39.2
   c. $0.392
   d. $392

18. Cut 12.85ft long stick into 0.5ft pieces. How many pieces will there be?

   a. 20.7
   b. 26.3
   c. 84.4
   d. 25.7

19. Which of the following has different quotient?

   a. 0.25 ÷ 0.5
   b. 2.5 ÷ 5
   c. 25 ÷ 5
   d. 0.025 ÷ 0.05

20. Sally ate 0.85 pounds of walnuts. Mary ate 0.258 pounds of walnuts. How many times as much walnuts Sally ate as Mary. Answer to the nearest tenth.

   a. 32
   b. 3.3
   c. 3.29
   d. 2.3

CHAPTER 2: FRACTIONS

2.0 Review Notes on Fractions

- A fraction is a number representing part of a whole. A fraction has the form, a/b, a is numerator and b is denominator. Ex. ¾ is a fraction, 3 is the numerator and 4 is the denominator. We read ¾ as three over four or three divided by four or three fourths.

- Proper fraction: the numerator is less than denominator.

- Improper fraction: the numerator is greater than or equal to the denominator.

- Mixed number: whole number combined with fraction.

- Dimensional analysis is a calculation involving measurement units as a factor

- Prime factorization of a number is writing the number as a multiple (product) of prime numbers. A prime number is a whole number greater than 1 that cannot be written as a product of two other whole numbers other than one and itself. Example: 5 is prime because 1 and 5 are the only factors of 5.

- Common factor is a number that is a factor of at least two numbers.

- The greatest common factor (GCF) of two (or more) numbers is the greatest number dividing all the numbers.

- Least Common Multiple (LCM) of two or more numbers is the smallest positive integer which is the multiple of the numbers.

- Equivalent fraction: a fraction obtained by multiplying (or dividing) the numerator and the denominator by the same nonzero number.

- To add/subtract two or more fractions with common denominator, add/subtract the numerators and divide by the common denominator.

- To add/subtract two (or more) fractions with different denominators:

    a. Find the LCM of the denominators.

    b. Convert each fraction to equivalent fraction using the LCM as a denominator of all terms.

    c. Add/subtract the numerators and divide by the common denominator which is the LCM.

- The reciprocal of a number is one divided by the number.

- To multiply fractions; multiply the numerators and divide by the product of the denominators.

- To divide two fractions; multiply the first fraction by the reciprocal of the second fraction.
- To add two (or more) mixed numbers, add the whole number parts and the fraction parts. Make sure to express the fractions using common denominator before adding or subtracting.
- To multiply two mixed numbers first convert the mixed numbers to improper fractions and multiply the resulting fractions.
- To Convert mixed number to fraction:
  - multiply the whole number in the mixed number by the denominator
  - add the result to the numerator of the fraction part (to get new numerator)
  - Put new numerator over the denominator
- To convert improper fraction to mixed number:
  - Divide the numerator by the denominator
  - Write down the whole **number** answer
  - Then write down any remainder above the denominator
- To compare fractions; convert each to equivalent fraction having the same denominator and compare the numerators
- Composite Number is a whole number that can be divided evenly by **numbers** other than 1 or itself.
- 1 is neither prime nor composite number.

2.1. Adding and Subtracting Fractions

2.1.1. Add & Subtract Fractions with Common Denominator

1. $\dfrac{3}{5} + \dfrac{2}{5} =$ _____

2. $\dfrac{13}{5} - \dfrac{2}{5} =$ _____

3. $\dfrac{32}{17} + \dfrac{2}{17} =$ _____

4. $\dfrac{7}{11} + \dfrac{19}{11} =$ _____

5. $\dfrac{13}{5} - \dfrac{13}{5} =$ _____

6. $\dfrac{28}{5} + \dfrac{2}{5} =$ _____

7. $\dfrac{3}{5} + \dfrac{3}{5} + \dfrac{2}{5} =$ _____

8. $\dfrac{22}{5} - \dfrac{3}{5} + \dfrac{2}{5} =$ _____

9. $\dfrac{22}{5} - \dfrac{3}{5} + \dfrac{2}{5} - \dfrac{21}{5} =$ _____

10. $\dfrac{22}{5} + \dfrac{3}{5} - \dfrac{2}{5} - \dfrac{21}{5} =$ _____

2.1.2 Add & Subtract Fractions with Different Denominator

1. $\dfrac{7}{3} - \dfrac{2}{5} =$ _____

2. $\dfrac{1}{7} + \dfrac{2}{5} =$ _____

3. $\dfrac{6}{8} + \dfrac{2}{5} =$ _____

4. $2 + \dfrac{3}{5} =$ _____

5. $\dfrac{11}{2} + \dfrac{12}{4} =$ _____

6. $\dfrac{19}{8} - \dfrac{2}{24} =$ _____

7. $6 - \dfrac{2}{7} =$ _____

8. $\dfrac{12}{14} - \dfrac{2}{5} =$ _____

9. $\dfrac{13}{5} - \dfrac{2}{6} =$ _____

10. $\dfrac{32}{14} - \dfrac{2}{7} =$ _____

11. $\dfrac{7}{10} + \dfrac{19}{15} =$ _____

12. $\dfrac{73}{25} - \dfrac{13}{5} =$ _____

13. $\dfrac{7}{2} + \dfrac{7}{3} + \dfrac{19}{6} =$ _____

14. $\dfrac{4}{3} + \dfrac{5}{6} + \dfrac{19}{5} =$ _____

15. $\dfrac{14}{3} - \dfrac{5}{6} + \dfrac{9}{5} =$ _____

16. $\dfrac{7}{8} + \dfrac{5}{6} - \dfrac{4}{5} =$ _____

17. $\dfrac{17}{28} + \dfrac{35}{14} + \dfrac{4}{5} =$ _____

18. $\dfrac{115}{38} + \dfrac{35}{19} + \dfrac{24}{3} =$ _____

19. $\dfrac{14}{28} + \dfrac{28}{14} + \dfrac{4}{8} =$ _____

20. $\dfrac{17}{28} + \dfrac{35}{14} - \dfrac{4}{5} =$ _____

2.2. Estimate the product of the numbers

1. $\dfrac{1}{8} \times \dfrac{1}{5} =$ _____

2. $\dfrac{7}{8} \times \dfrac{2}{5} =$ _____

3. $\dfrac{1}{3} \times \dfrac{4}{5} =$ _____

4. $\dfrac{1}{3} \times 11 =$ _____

5. $\dfrac{2}{5} \times 26 =$ _____

6. $\dfrac{4}{9} \times \dfrac{9}{10} =$ _____

7. $\dfrac{7}{8} \times \dfrac{9}{8} =$ _____

8. $\dfrac{5}{6} \times \dfrac{1}{11} =$ _____

9. $\dfrac{13}{3} \times \dfrac{11}{4} =$ _____

10. $24 \times \dfrac{2}{5} =$ _____

2.3. Proper and Improper Fractions

Identify whether the fraction is proper or improper

a. two – tenths = _____

b. $\dfrac{3}{11} =$ _____

c. $\dfrac{33}{11} =$ _____

d. $\dfrac{23}{71} =$ _____

e. $3\dfrac{2}{9} =$ _____

f. $\dfrac{38}{3} = $ _____

g. twenty four – seventeenths = _____

h. thirty four – hundredths = _____

i. $\dfrac{31}{32} = $ _____

j. $\dfrac{204}{365} = $ _____

2.4. Mixed to Improper Fraction

1. Convert each mixed number to an improper fraction

1. $5\dfrac{3}{20} = $ _____

2. $20\dfrac{1}{5} = $ _____

3. $5\dfrac{2}{25} = $ _____

4. $5\dfrac{22}{25} = $ _____

5. $1\dfrac{7}{10} = $ _____

6. $2\dfrac{2}{100} = $ _____

7. $8\dfrac{1}{5} = $ _____

8. $1\dfrac{1}{4} = $ _____

9. $5\dfrac{3}{4} = $ _____

10. $7\dfrac{5}{11} = $ _____

11. $3\dfrac{4}{33} = $ _____

12. $6\dfrac{13}{213} = $ _____

13. $2\dfrac{3}{21} = $ _____

14. $7\dfrac{1}{11} = $ _____

15. $21\dfrac{7}{8} = $ _____

16. $5\dfrac{22}{33} = $ _____

17. $8\dfrac{1}{22} = $ _____

18. $75\dfrac{5}{110} = $ _____

19. $4\dfrac{1}{9} = $ _____

20. $312\dfrac{1}{3} = $ _____

2.5. Improper Fraction to Mixed Number

Convert to mixed number

1. $\dfrac{5}{4} =$ _____

2. $\dfrac{41}{20} =$ _____

3. $\dfrac{11}{3} =$ _____

4. $\dfrac{25}{21} =$ _____

5. $\dfrac{47}{22} =$ _____

6. $\dfrac{33}{27} =$ _____

7. $\dfrac{78}{9} =$ _____

8. $\dfrac{58}{47} =$ _____

9. $\dfrac{35}{34} =$ _____

10. $\dfrac{204}{15} =$ _____

11. $\dfrac{300}{17} =$ _____

12. $\dfrac{604}{27} =$ _____

13. $\dfrac{384}{25} =$ _____

14. $\dfrac{1023}{200} =$ _____

15. $\dfrac{425}{20} =$ _____

16. $\dfrac{890}{21} =$ _____

2.6. Add & Subtract Mixed Numbers

2.6.1 Add or Subtract Fractions with Common Denominator

1. $2\dfrac{5}{6} + 7\dfrac{2}{6} =$ _____

2. $4\dfrac{5}{13} + 7\dfrac{1}{13} =$ _____

3. $9\dfrac{5}{7} - 7\dfrac{3}{7} =$ _____

4. $7\dfrac{2}{6} - 3\dfrac{1}{6} =$ _____

5. $8\dfrac{2}{6} - 3\dfrac{5}{6} =$ _____

6. $5\dfrac{8}{12} + 7\dfrac{5}{12} =$ _____

7. $12\dfrac{8}{15} - 7\dfrac{9}{15} =$ _____

8. $8\dfrac{8}{11} + \dfrac{5}{11} =$ _____

9. $6\dfrac{1}{7} - \dfrac{3}{7} =$ _____

10. $12\frac{2}{7} + 13\frac{5}{7} =$ _____

2.6.2. Add & Subtract with Different Denominator

11. $13\frac{5}{3} + 7\frac{2}{6} =$ _____

12. $4\frac{5}{3} - 2\frac{1}{4} =$ _____

13. $7\frac{5}{6} - 3\frac{3}{7} =$ _____

14. $11\frac{2}{5} - 7\frac{1}{6} =$ _____

15. $9\frac{2}{6} - 4\frac{5}{10} =$ _____

16. $15\frac{1}{10} + 7\frac{5}{12} =$ _____

17. $13\frac{8}{10} - 7\frac{9}{15} =$ _____

18. $9\frac{8}{11} + 1\frac{5}{10} =$ _____

19. $26\frac{1}{3} - 6\frac{4}{7} =$ _____

20. $7\frac{2}{12} - 2\frac{5}{7} =$ _____

2.7 Multiply Fractions

2.7.1 Multiply fraction by whole number

1. $4 \times \frac{7}{4} =$ _____

2. $13 \times \frac{5}{13} =$ _____

3. $1 \times \frac{5}{4} =$ _____

4. $18 \times \frac{2}{9} =$ _____

5. $24 \times \frac{7}{6} =$ _____

6. $63 \times \frac{5}{3} =$ _____

7. $88 \times \frac{24}{8} =$ _____

8. $55 \times \frac{11}{6} =$ _____

9. $120 \times \dfrac{15}{10} =$ _____

10. $213 \times \dfrac{421}{27} =$ _____

11. $3 \times \dfrac{5}{3} \times 10 =$ _____

12. $8 \times 43 \times 0 \times \dfrac{215}{3} =$ _____

13. $111 \times \dfrac{6}{5} \times 2 =$ _____

14. $213 \times \dfrac{1675}{29} =$ _____

15. $100 \times \dfrac{200}{20} \times 40 =$ _____

2. 7. 2 Multiply fraction by fraction

1. $\dfrac{1}{3} \times \dfrac{2}{5} =$ _____

2. $\dfrac{3}{4} \times \dfrac{3}{5} =$ _____

3. $\dfrac{11}{2} \times \dfrac{12}{4} =$ _____

4. $\dfrac{19}{8} \times \dfrac{2}{24} =$ _____

5. $\dfrac{3}{7} \times \dfrac{2}{7} =$ _____

6. $\dfrac{2}{14} \times \dfrac{2}{5} =$ _____

7. $\dfrac{13}{5} \times \dfrac{2}{6} =$ _____

8. $\dfrac{32}{14} \times \dfrac{2}{7} =$ _____

9. $\dfrac{7}{10} \times \dfrac{19}{15} =$ _____

10. $\dfrac{13}{25} \times \dfrac{13}{5} = $ _____

11. $\dfrac{7}{10} \times \dfrac{7}{25} \times \dfrac{19}{15} = $ _____

12. $\dfrac{4}{3} \times \dfrac{5}{6} \times \dfrac{19}{5} = $ _____

13. $\dfrac{14}{3} \times \dfrac{5}{6} \times \dfrac{9}{5} = $ _____

14. $\dfrac{7}{8} \times \dfrac{5}{6} \times \dfrac{4}{5} = $ _____

15. $\dfrac{17}{28} \times \dfrac{35}{14} \times \dfrac{4}{5} = $ _____

16. $\dfrac{115}{38} \times \dfrac{35}{19} \times \dfrac{24}{3} = $ _____

17. $\dfrac{17}{28} \times \dfrac{35}{14} \times \dfrac{0}{5} = $ _____

2.8. Multiplying Mixed Numbers

Multiply:

1. $5\frac{2}{6} \times 3\frac{1}{4} =$ _____

2. $10\frac{2}{10} \times 4\frac{5}{10} =$ _____

3. $15\frac{1}{10} \times 7\frac{5}{12} =$ _____

4. $13\frac{8}{10} \times 7\frac{9}{15} =$ _____

5. $9\frac{8}{11} \times 1\frac{5}{10} =$ _____

6. $2\frac{1}{3} \times 6\frac{4}{7} =$ _____

7. $4\frac{5}{3} \times 2\frac{1}{4} =$ _____

8. $13\frac{5}{3} \times 7\frac{2}{6} =$ _____

9. $3\frac{5}{4} \times 9\frac{2}{6} =$ _____

10. $1\dfrac{5}{40} \times 7\dfrac{2}{6} =$ _____

2.9. Dimensional analysis

2.9.1 Convert to the given unit of length (answer round to two decimal place)

Notations: ml. = mile, ft. = feet, yd. = yard, and in. = inch.

1. 7 ft. = _____ in.

2. 4.5 yd. = _____ in

3. 2.5 miles. = _____ yd.

4. 38 in. = _____ in.

5. 2.5 ft. = _____ yd.

6. 72 in. = _____ yd.

7. 7 ml. = _____ ft.

8. 24 ml. = _____ yd.

9. 2,640 ft. = _____ ml.

10. 3,520 yd. = _____ in.

11. 10,560 yd. = _____ml.

12. 24 in. = _____ ft.

13. 12 ft. =_____ml.

14. 20 yd. = _____ml.

15. 72 in. = _____ ft.

16. 0.25 ml = _____ ft.

17. 3,600 in. = _____ yd.

18. 4,600 ft. = _____ml.

19. 20,000 in. = _____ml.

20. 3,520 yd. = _____ ml.

2.9.2. Convert to the given capacity (pt. = pint, c = cup, gal. = gallon, qt. = quart)

1. 7.5 pt. _____ C

2. 4.5 gal _____ qt.

3. 36 gal _____ qt.

4. 28 qt. _____ gal

5. 28 oz. _____ C

6. 6 C _____ pt.

7. 16 C _____ pt.

8. 8 C _____ oz.

9. 8 qt. _____ gal

10. 7 gal _____ qt.

11. 12 qt _____ gal

12. 6 pt. _____ C

13. 3 C _____ oz.

14. 16 C _____ oz.

15. 5 C _____ oz.

16. 14 C _____ pt.

2.10. Divide Fractions

2.10.1 Dividing a whole number by fraction.

1. $16 \div \dfrac{4}{5} =$ _____

2. $36 \div \dfrac{6}{13} =$ _____

3. $34 \div \dfrac{51}{4} =$ _____

4. $8 \div \dfrac{2}{9} =$ _____

5. $12 \div \dfrac{7}{6} =$ _____

6. $36 \div \dfrac{9}{4} =$ _____

7. $33 \div \dfrac{3}{6} =$ _____

8. $55 \div \dfrac{11}{6} =$ _____

9. $120 \div \dfrac{10}{6} =$ _____

10. $13 \div \dfrac{42}{27} =$ _____

11. $15 \div \dfrac{5}{7} =$ _____

12. $0 \div \dfrac{215}{3} =$ _____

13. $121 \div \dfrac{11}{5} =$ _____

14. $222 \div \dfrac{111}{29} =$ _____

15. $200 \div \dfrac{200}{20} =$ _____

2.10.2. Dividing fraction by fraction.

1. $\dfrac{0}{20} \div \dfrac{2}{5} =$ _____

2. $\dfrac{9}{5} \div \dfrac{81}{10} =$ _____

3. $\dfrac{11}{2} \div \dfrac{12}{4} =$ _____

4. $\dfrac{19}{8} \div \dfrac{2}{24} =$ _____

5. $\dfrac{3}{7} \div \dfrac{3}{7} =$ _____

6. $\dfrac{32}{14} \div \dfrac{2}{5} =$ _____

7. $\dfrac{81}{5} \div \dfrac{9}{60} =$ _____

8. $\dfrac{64}{14} \div \dfrac{2}{7} =$ _____

9. $\dfrac{27}{10} \div \dfrac{54}{15} =$ _____

10. $\dfrac{13}{25} \div \dfrac{13}{25} =$ _____

11. $\dfrac{7}{10} \div \dfrac{19}{15} =$ _____

12. $\dfrac{4}{3} \div \dfrac{5}{6} \div \dfrac{19}{5} =$ _____

13. $\dfrac{10}{3} \div \dfrac{5}{6} \div \dfrac{9}{5} =$ _____

2.11. Dividing Mixed Numbers

Divide

1. $5\dfrac{5}{3} \div 7\dfrac{2}{6} =$ _____

2. $13\dfrac{5}{3} \div 2\dfrac{1}{4} =$ _____

3. $7\dfrac{5}{6} \div 3\dfrac{3}{7} =$ _____

4. $11\dfrac{2}{5} \div 7\dfrac{1}{6} =$ _____

5. $9\dfrac{2}{6} \div 4\dfrac{5}{10} =$ _____

6. $5\dfrac{1}{10} \div 7\dfrac{5}{12} =$ _____

7. $3\dfrac{8}{10} \div 7\dfrac{9}{15} =$ _____

8. $9\dfrac{8}{11} \div 1\dfrac{5}{10} =$ _____

9. $6\dfrac{1}{3} \div 6\dfrac{4}{7} =$ _____

10. $5\dfrac{2}{12} \div 2\dfrac{5}{7} =$ _____

2.12. Prime Factorization

2.12.1 Prime and Composite numbers

1. List all prime numbers between 1 and 30

   Answer = _____

2. List all composite numbers between 2 and 35

   Answer = _____

3. Write the following numbers as a product of prime numbers

a. 12 = _____

b. 9 = _____

c. 18 = _____

d. 16 = _____

e. 34 = _____

f. 75 = _____

2.12.2. List all the factors of the number.

1. 8 = _____

2. 11 = _____

3. 12 = _____

4. 28 = _____

5. 52 = _____

6. 102 = _____

7. 68 = _____

8. 210 = _____

9. 17 = _____

10. 59 = _____

11. 43 = _____

12. 51 = _____

13. 129 = _____

14. 75 = _____

15. 150 = _____

16. 120 = _____

2.13. Greatest Common Factor

Find the greatest common factor of each question

1. 7, 10 = _____

2. 12, 30 = _____

3. 3, 11 = _____

4. 50, 20 = _____

5. 36, 90 = _____

6. 37, 23 = _____

7. 81, 72 = _____

8. 38, 98 = _____

9. 100, 200 = _____

10. 12, 120 = _____

11. 360, 96 = _____

12. 128, 144 = _____

13. 378, 462 = _____

14. 8, 24, 36 = _____

15. 31, 1 = _____

16. 27, 90 = _____

2.14. Least Common Multiple

Find the Least Common Multiple

1. 11, 10 = _____

2. 5, 6, and 8 = _____

3. 16, 20 = _____

4. 12, 16 = _____

5. 18, 60 = _____

6. 6, 10, 12 = _____

7. 4, 12, 16, 24 = _____

8. 6, 8, 10 = _____

9. 10, 30, 40 = _____

10. 80, 360 = _____

11. 216, 144 = _____

12. 136, 72 = _____

13. 20, 6 = _____

14. 28, 12 = _____

15. 30, 12 = _____

16. 1, 30 = _____

2.15. Simplifying Fractions

1. $\dfrac{8}{72} =$ _____

2. $\dfrac{3}{18} =$ _____

3. $\dfrac{28}{34} =$ _____

4. $\dfrac{24}{48} =$ _____

5. $\dfrac{51}{3} =$ _____

6. $\dfrac{8}{56} =$ _____

7. $\dfrac{22}{66} =$ _____

8. $\dfrac{12}{108} =$ _____

9. $\dfrac{60}{114} =$ _____

10. $\dfrac{68}{17} =$ _____

11. $\dfrac{45}{9} =$ _____

12. $\dfrac{20}{35} =$ _____

13. $\dfrac{30}{35} =$ _____

14. $\dfrac{50}{25} =$ _____

15. $\dfrac{21}{48} =$ _____

16. $\dfrac{20}{48} =$ _____

17. $\dfrac{45}{36} =$ _____

18. $\dfrac{20}{120} =$ _____

19. $\dfrac{32}{160} =$ _____

20. $\dfrac{60}{160} =$ _____

2.16 Simplify and Reduce Each Fraction to Lowest Form

1. $\dfrac{10}{125} =$ _____

2. $\dfrac{55}{20} =$ _____

3. $\dfrac{140}{20} =$ _____

4. $\dfrac{448}{24} =$ _____

5. $\dfrac{240}{1200} =$ _____

6. $\dfrac{550}{55} =$ _____

7. $\dfrac{90}{810} =$ _____

8. $\dfrac{243}{28} =$ _____

9. $\dfrac{500}{2500} =$ _____

10. $\dfrac{216}{360} =$ _____

11. $\dfrac{216}{360} \times \dfrac{4}{6} =$ _____

12. $\frac{16}{360} \times \frac{6}{48} =$ _____

13. $\frac{200}{60} \times \frac{40}{60} =$ _____

14. $\frac{7}{8} \div \frac{28}{24} =$ _____

15. $\frac{36}{216} \div \frac{4}{16} =$ _____

16. $\frac{36}{216} \div \frac{4}{16} \times \frac{1}{24} =$ _____

17. $\frac{81}{216} \div \frac{18}{36} \times \frac{6}{1} =$ _____

18. $\frac{54}{27} \div \frac{6}{81} =$ _____

19. $\dfrac{36}{216} \div \dfrac{4}{16} =$ _____

20. $\dfrac{25}{100} \div \dfrac{4}{8} =$ _____

2.17. Mixed numbers with mixed operations

1. $8\dfrac{2}{12} \div 2\dfrac{5}{7} - 1\dfrac{5}{7} =$ _____

2. $6\dfrac{8}{11} \times 3\dfrac{5}{10} - \dfrac{2500}{110} =$ _____

3. $15\dfrac{1}{10} + 7\dfrac{5}{12} \times \dfrac{1}{12} =$ _____

4. $3\frac{3}{5} - 2\frac{2}{3} + 4 \times \frac{1}{12} =$ _____

5. $6\frac{3}{5} \div 2\frac{2}{3} + 5 \times \frac{1}{12} =$ _____

2.18. Ordering Fractions

2.18.1. Use < or > or = to compare the following fractions

1. $\frac{13}{4} \;\;\; \frac{2}{3}$

2. $\frac{4}{7} \;\;\; \frac{3}{7}$

3. $\frac{12}{5} \;\;\; \frac{2}{3}$

4. $\frac{10}{18} \;\;\; \frac{5}{9}$

5. $\dfrac{1}{3} - \dfrac{3}{9}$

6. $\dfrac{4}{5} - \dfrac{12}{41}$

7. $\dfrac{23}{46} - \dfrac{30}{4}$

8. $\dfrac{12}{32} - \dfrac{9}{18}$

9. $\dfrac{7}{24} - \dfrac{7}{15}$

10. $\dfrac{7}{8} - \dfrac{5}{16}$

11. $\dfrac{3}{5} - \dfrac{16}{24}$

12. $\dfrac{13}{12} - \dfrac{24}{144}$

13. $\dfrac{25}{125} - \dfrac{2}{10}$

14. $4\dfrac{5}{12} - 3\dfrac{5}{13}$

15. $3\dfrac{14}{13} - \dfrac{32}{8}$

16. $2\dfrac{5}{3} - \dfrac{65}{13}$

17. $2\dfrac{5}{13} - 0\dfrac{5}{1}$

18. $4\dfrac{5}{13} + 5\dfrac{8}{13} - 10$

19. $7\dfrac{2}{3} - \dfrac{25}{3}$

20. $1\dfrac{5}{7} - \dfrac{5}{13} - 0$

21. $9\dfrac{8}{11} \div 1\dfrac{5}{10} - 10\dfrac{8}{11}$

22. $2\dfrac{8}{11} \times 1\dfrac{5}{10} - \dfrac{8}{11} \div \dfrac{5}{10}$

2.18.2 Order the numbers from the least to the greatest

1. $\dfrac{1}{10}, \dfrac{2}{5}, \dfrac{1}{2} =$ _____

2. $\dfrac{6}{10}, \dfrac{2}{3}, \dfrac{1}{2} =$ _____

3. $\dfrac{7}{4}, \dfrac{1}{10}, \dfrac{2}{5}, \dfrac{1}{2} =$ _____

4. $\dfrac{4}{5}, \dfrac{5}{6}, \dfrac{6}{7}, \dfrac{1}{2} =$ _____

5. $\dfrac{7}{11}, \dfrac{7}{4}, \dfrac{1}{10}, \dfrac{2}{5}, \dfrac{13}{12} =$ _____

6. $\dfrac{30}{4}, \dfrac{14}{5}, \dfrac{25}{6}, \dfrac{64}{7}, \dfrac{10}{2} =$ _____

7. $3\dfrac{3}{4}, 1\dfrac{4}{5}, 2\dfrac{5}{6} =$ _____

8. $10\dfrac{1}{4}, 10\dfrac{2}{5}, 6\dfrac{5}{6}, 9\dfrac{7}{5}, \dfrac{1}{2} =$ _____

9. $1\dfrac{11}{5}, 2\dfrac{2}{5}, \dfrac{15}{6}, 3\dfrac{6}{5}, 7\dfrac{1}{2} =$ _____

10. $\dfrac{18}{7}, \dfrac{200}{56}, \dfrac{215}{36}, 23\dfrac{6}{5} =$ _____

11. $1\dfrac{11}{5}, 2\dfrac{2}{5}, \dfrac{15}{16}, \dfrac{6}{5}, 10 =$ _____

12. $\dfrac{101}{51}, 2\dfrac{2}{51}, 1\dfrac{17}{16}, 2\dfrac{6}{5}, 5 =$ _____

2.19. Word problems involving fractions

1. Bolt has 48 dollars which is 2.4 as much as his sister Jenny. How much money does Jenny have?

   Answer = _____

2. How many 2/5 in. pieces of ropes can be made from 2ft. long rope?

   Answer = _____

3. John goes to church every 3 days. Luke goes to church every 4 days. John and Luke met for the first time on June 2, 2020. When will they meet for the second time?

   Answer = _____

4. Mike has 32 red, 40 blue, and 48 green marbles. He wants to put an equal number of red, blue, and green marbles into a box. What is the greatest number of boxes that can be made so that no marble is left?

   Answer = _____

5. A chef is creating individual servings of starters. He has 6 bread sticks and 8 carrots. If he wants each serving to be identical, with no food left over, what is the greatest number of servings the chef can create?

6. Mark has 1/3 of a dollar. He used ¼ of his money to buy a pencil. What fraction of the dollar is used for his pencil?

   Answer = _____

7. Ms. Lora has 3/5 of a pound of milk powder. She used 5/7 of what she had to make a glass of juice. What fraction of a pound of milk did Ms. Lora use to make the juice?

   Answer = _____

8. ¼ of a rectangular wall is to be painted green. Two-fifth of the green portion is to be reserved for pictures. What fraction of the wall is reserved for the pictures?

   Answer = _____

9. Tom and nine friends earned $180 washing cars. Each of them received 1/10 of the total amount. How much money did each earn?

Answer = _____

10. If you drink 45/6 liter of water in two and half days. How much water will you drink in a day?

Answer = _____

11. There are 10 boxes of books. Three-fifth of the box is filled with nonfiction books. How many nonfiction books are there in all boxes?

Answer = _____

12. How many pizzas will you buy to feed 32 friends if each eats 3/8 of a pizza?

Answer = _____

13. To make a dozen cookies you need 1/20 of a pound of sugar, how many dozens of cookies will you make if you use 200 pounds of sugar?

Answer = _____

14. Divide 27 ½ dollars among three friends. How much does each get?

Answer = _____

15. Thirty-two marbles are divided evenly among 4 boys. How many marbles does each boy get?

Answer = _____

16. Sixty books are given to three students. What fraction of the books does each student receive?

Answer = _____

17. What is one – fourth of 80 ½?

Answer = _____

18. It rained the same amount for 3 days. The total amount of rain received was 10/3 inches. Find the amount of rain received every day.

    Answer = _____

19. A rope is cut into 3 pieces of equal length. How long is each piece if the length of the rope was 90 ½ ft.?

    Answer = _____

20. Five students are participating in a relay race that is 2 1/3 miles long. If each student runs the same distance, how far does each student run?

    Answer = _____

21. Kimberly takes 3/2 hours to type 7 pages. How long will it take her to type the book titled, United States History which has 455 pages?

    Answer = _____

22. Sam's dad allows Sam to eat 2 ½ chocolate bars every week. How many chocolate bars will Sam eat in a year?

Answer = _____

23. A circle is divided into six equal sections. If each section is cut in fourth. How many pieces will the circle have?

   Answer = _____

24. At Six flags John used four coins for each game. John had 104 coins. How many games did he play?

   Answer = _____

25. A mailman distributes 166 mails in an hour. If the mailman works for 32 and half hours. How many mails will he deliver?

   Answer = _____

26. Jenny is 17/3 ft. tall. Her younger sister is 13/3 ft. How taller is Jenny than her younger sister?

Answer = _____

27. Your seven friends and you want to spend $3200 on vacation. Will there be some money left over if the money is distributed equally?

Answer = _____

28. Kimberly was putting pencils into pencil cases. She had 7 pencil cases that would hold 12 pencils each. She had 84 pencils. Would all pencils fit in the 7 cases?

Answer = _____

29. Tom has 84 crayons and he would like to give these crayons to his six friends. How many crayons will each friend get?

Answer = _____

30. There are 48 students in Markos' class. Markos assigned three different questions to each student. How many problems did Markos bring to the class?

Answer = _____

31. Alex wants to distribute 56 candies to 7 children. How many candies will each child get?

   Answer = _____

32. Three pumpkin pies are divided evenly among 8 people. How much pie did each person get?

   Answer = _____

33. Six chocolate bars are shared equally among 8 children. Each child shared his/her share among 8 other children. What part of a chocolate bar does each child receive?

   Answer = _____

34. A builder needs 8 trees. He cuts each tree into sixth pieces. How many pieces of does he have?

   Answer = _____

35. A bus could carry one-eighth of the passengers, a train carries. If the bus has 25 passengers, how many passengers would the train carry?

   Answer = _____

36. How many bottles would it take to fill up a 24liter bucket if a bottle of milk is four-fifth of a liter?

   Answer = _____

37. An athlete runs around a rectangular soccer field that has a perimeter of 400 meters. If the athlete runs 12,200 meters, will he complete at least 20 rounds?

   Answer = _____

38. How many one-fourth cheese sticks are in a plastic bag holding 10 cheese sticks?

   Answer = _____

39. A father is dividing up one-third of his money to his 3 children. The second and third children get twice as much as the first child. What fraction of the money did the first child get?

   Answer = _____

40. A 5th grade project has 110 questions. How long will it take to complete these questions if a student can do one-tenth of the questions each day?

   Answer = _____

41. What is one –sixth of one –fifth of one?

   Answer = _____

42. Papa John pizza uses one-sixth of a box of onions a day. How many days would 10 full boxes of onions last?

   Answer = _____

43. A bottle of water weighs one- eighth of a jug of water. If you have three full jugs of water, how many persons can be served if each person is drinking 2/3 of a bottle of water?

   Answer = _____

44. A family had 20 pounds of meat. How many days would it take the family to use the 20 pounds of meat if the family needs one-third of a pound a day?

   Answer = _____

45. A bag of chips is distributed among 5 friends. Each share to three other friends. What fraction of the bag does each get?

   Answer = _____

2. 20. Chapter 2 Test

1. If you cut a 42 ft. rope into pieces, each piece 3/2 ft. long. How many pieces will you have?

   a. 24
   b. 25
   c. 26
   d. 28

2. Sam and his friends had lunch at Olive Garden restaurant. The reasonable tip is 18%. If the bill is $64. How much money are they expected to pay for the tip?

   a. $12.52
   b. $11.52
   c. $10.52
   d. $9.52

3. $3\frac{3}{5} - 2\frac{2}{3} + 4 \times \frac{1}{12} \times 0 =$ _____

   a. 1
   b. 14/15
   c. 1.5
   d. 2

4. The least common multiple of the numbers 18 and 24 is _____.

   a. 72
   b. 36
   c. 108
   d. 6

5. Which of the following is not a factor of 18?

   a. 2
   b. 3
   c. 8
   d. 9

6. $5\frac{2}{3} - 4\frac{2}{5} =$ _____

   a. $1\frac{4}{5}$
   b. 1
   c. $1\frac{4}{15}$

64

d. No answer

7. Which of the following is not equivalent to $\dfrac{3}{7}$?

   a. $\dfrac{30}{70}$

   b. $\dfrac{9}{21}$

   c. $\dfrac{18}{79}$

   d. $\dfrac{6}{14}$

8. A fifth grader needs 1/3 of an hour to take a shower and ¼ of an hour to eat breakfast. What fraction of an hour does the student need altogether?

   a. $\dfrac{3}{7}$

   b. $\dfrac{1}{2}$

   c. $\dfrac{7}{12}$

   d. $\dfrac{1}{7}$

9. Change $7\dfrac{3}{15}$ to improper fraction.

   a. $\dfrac{108}{15}$

   b. $\dfrac{118}{15}$

   c. $\dfrac{10}{15}$

   d. $\dfrac{21}{15}$

10. Change $\dfrac{128}{12}$ to mixed number

a. $12\dfrac{1}{12}$

b. $10\dfrac{8}{12}$

c. $100\dfrac{28}{12}$

d. 12

11. Multiply: $5 \times 4\dfrac{2}{5}$

    a. 22

    b. $\dfrac{110}{125}$

    c. $100\dfrac{28}{12}$

    d. $\dfrac{22}{12}$

12. Subtract: $\dfrac{14}{3} - \dfrac{5}{6}$

    a. $3\dfrac{5}{6}$

    b. $\dfrac{9}{6}$

    c. $\dfrac{9}{3}$

    d. $\dfrac{22}{12}$

    e. No answer

13. Divide: $3 \div \dfrac{2}{9}$

    a. $\dfrac{6}{9}$

    b. $\dfrac{27}{2}$

c. $\dfrac{2}{3}$

d. $\dfrac{2}{27}$

14. Divide $2\dfrac{5}{6} \div 3\dfrac{3}{7}$

a. $\dfrac{16}{42}$

b. $\dfrac{68}{7}$

c. $\dfrac{408}{40}$

d. $\dfrac{119}{144}$

15. The order of the numbers: $\dfrac{6}{10}, \dfrac{2}{3}, \dfrac{1}{2}$ is

    a. From least to greatest

    b. From greatest to least

    c. There is no order

16. Order the fractions from least to greatest 1/6, 3/5, 7/6, 1/3, 4/9.

    a. 1/6, 1/3, 4/9, 3/5, 7/6

    b. 1/6, 4/9, 1/3, 3/5, 7/6

    c. 1/6, 1/3, 4/9, 7/6, 3/5

    d. 1/6, 1/3, 3/5, 4/9, 7/6

17. Mark ate 1/4 of an apple on Monday and 2/3 on Tuesday. How much more apple did Mark eat on Tuesday than Monday?

    a. $\dfrac{1}{4}$

    b. $\dfrac{7}{4}$

c. $\dfrac{11}{4}$

d. $\dfrac{5}{12}$

18. 1/5 > 3/18?

   a. True

   b. False

   c. No answer

19. A third of 3 is

   a. 9

   b. 3

   c. 1

   d. 6

20. How many hours are there in 250 minutes?

   a. 4 hours

   b. $4\dfrac{4}{8}$ hours

   c. $4\dfrac{1}{6}$ hours

   d. 6 hours

CHAPTER 3: RATIOS, PERCENTS & PROPORTIONS

3.0 Review Notes on Ratios, Percent, and Proportions

- Ratio: comparison of two quantities in which order is important.
- Rate: ratio comparing two quantities with different units

- Unit rate: ratio with denominator one.

- Proportion: connecting two ratios with equal sign

- Percent Proportion: has two equivalent ratios, with one ratio (a part to the whole) and the other ratio with a denominator of 100.

- Equivalent ratios: two or more ratios that are proportional.

- To find an equivalent ratio multiply both the numerate and denominator by the same number.

To convert a fraction to a percent:

- Divide the numerator of the fraction by the denominator
- Multiply by 100 (Move the decimal point two places to the right)
- Round the answer to the desired precision.
- Follow the answer with the % sign

To convert Percent to Fraction:

- Drop the percent sign from the number
- Divide the number by 100

To convert decimal to percent:

- Multiply the decimal by 100
- Add a percent sign after the answer

To convert percent to decimal

- Drop the percent sign from the number
- Divide the number by 100 and write the number in decimal form
- Example: Converting 25% to decimal = 25/100 = 0.25

To solve proportion

- Assign letter for the unknown
- Cross multiply
- Divide the result of the product of the numbers by the number attached to the unknown

3.1. Ratio

3.1.1. Write the ratio specifying the type of ratio when necessary.

1. The ratio of pens to pencils in a class is 14: 29. For every _____ pencils there are _____ pens.

2. In a park for every 6 boys there were 9 girls. The ratio of girls to boys is _____.

3. At the store the ratio of math books sold to science books sold was 14:17. For every _____ science books sold there were _____ math books sold.

4. The ratio of oranges to apples in a grocery is 5/7. For every 49 apples there are _____ oranges.

5. Ruth has seven pencils and 9 pens. Ratio of pencils to pens _____.

6. Of 26 teachers, 14 are males. Ratio of males to all _____.

7. Living in a house is preferred over living in an apartment by 8 of 10 residents_____.

8. Human nose to eyes_____.

9. In a house, there are 16 windows and 4 doors. Windows to doors _____.

10. At a restaurant, there are 64 chairs and 8 tables, Chairs to tables _____.

11. There are 3 baseball fields and 7 basketball courts, Baseball to basketball _____.

12. Of 1000 travelers, 750 drive cars and 250 take airplanes. Cars to airplanes_____.

13. A sample of 100 students asked what their favorite subject is, 40 said language arts, 35 said science and 25 said math. Write three possible ratios

   Answer = _____

   Answer = _____

   Answer = _____

14. Seven- tenths of the people entering the movie theater are younger than 25 years old___.

15. Write the fraction representing the ratio 5 to 7_____.

16. Find the ratio A to B, A to C, and B to C where A = 2, B = 3, C = 5

   Answer =_____, _____, and _____.

3.1. 2. More on Ratios

1. The table shows a sample of students' favorite sports.

Sport	Number of students
Football	20
Soccer	30
Basketball	40
Tennis	10

Answer the following question based on the information given in the table

a. Ratio of students who love soccer to basketball _____.

b. Ratio of students who love tennis to basketball _____.

c. Ratio of students who love tennis to football_____.

d. Ratio of students who love football to all sports_____.

e. Ratio of students who love basketball to soccer _____.

f. Ratio of students who love tennis to all sports _____.

g. Ratio of students who love soccer to tennis _____.

h. Ratio of students who love all sports to basketball _____.

2. Write each ratio as a fraction in lowest terms

   a. 15 inches to 60 inches = _____

   b. $25 to $ 125 = _____

   c. 40 seconds 60 seconds = _____

   d. 150 miles to 7500 miles = _____

   e. 200 liters to 50 liters = _____

f. 36 girls to 108 boys = _____

g. 17 oranges to 51 apples = _____

h. $21 for 700 candies = _____

i. 25 wrong answers out of 100 questions = _____

j. 250 miles per 25 gallons = _____

k. 6 tables for 48 people = _____

3.2 Rates and Unit rates

1. Find the unit rate if the ratio is 6 to 3 _____

2. Laura runs 24 miles in 3 hours. Find the rate_____

3. In question 2, what is the unit rate? _____

4. Bob 25 miles in 5 minutes. What is the unit rate? _____

5. What is the unit rate if 720 miles is traveled on 24 gallons of gasoline? _____

6. The unit rate if an insect travels 20 feet in 4 seconds _____

7. If you drive 240 miles in 6 hours, how far will you go in an hour? _____

8. Rate of 200 meters per 10 seconds, has a unit rate of _____

9. Sam wrote 63 pages in 7 hours. How many pages did he write per hour? _____

10. Nancy read 64 pages in 4 hours. Sally read 48 pages in 2 5 hours. Who read more pages per hour? _____

3.3. Equivalent ratio and Proportions

3.3.1. Find two equivalent fractions for each fraction

1. $\dfrac{2}{5} =$ _____

2. $\dfrac{3}{4} =$ _____

3. $\dfrac{5}{6} =$ _____

4. $\dfrac{11}{7} =$ _____

5. $\dfrac{3}{12} =$ _____

6. $\dfrac{2}{19} =$ _____

7. $\dfrac{1}{13} =$ _____

8. $\dfrac{5}{17} =$ _____

9. $\dfrac{3}{7} =$ _____

10. $\dfrac{50}{234} = $ _____

11. $\dfrac{31}{70} = $ _____

12. $\dfrac{29}{67} = $ _____

3.3.2. Complete to make the fraction proportional

1. $\dfrac{1}{3} = \dfrac{6}{?} = $ _____

2. $\dfrac{1}{4} = \dfrac{?}{20} = $ _____

3. $\dfrac{?}{28} = \dfrac{7}{4} = $ _____

4. $\dfrac{11}{?} = \dfrac{1}{43} = $ _____

5. $\dfrac{9}{12} = \dfrac{81}{?} = $ _____

6. $\dfrac{18}{?} = \dfrac{90}{125} = $ _____

7. $\dfrac{14}{21} = \dfrac{?}{3} = $ _____

8. $\dfrac{100}{400} = \dfrac{1}{?} = $ _____

9. $\dfrac{4}{18} = \dfrac{2}{?} = $ _____

10. $\dfrac{1}{?} = \dfrac{9}{90} = $ _____

11. $\frac{7}{4} = \frac{14}{?} =$ _____

12. $\frac{?}{4} = \frac{9}{4} =$ _____

13. $\frac{1}{14} = \frac{24}{?} =$ _____

14. $\frac{14}{10} = \frac{?}{5} =$ _____

15. $\frac{17}{411} = \frac{?}{822} =$ _____

16. $\frac{1}{4} = \frac{?}{4} =$ _____

17. $\frac{12}{18} = \frac{24}{?} =$ _____

18. $\frac{9}{72} = \frac{18}{?} =$ _____

19. $\frac{?}{5} = \frac{12}{60} =$ _____

20. $\frac{5}{30} = \frac{45}{?} =$ _____

21. $\frac{?}{5} = \frac{18}{5} =$ _____

22. $\frac{13}{5} = \frac{26}{?} =$ _____

23. $\frac{11}{44} = \frac{?}{88} =$ _____

24. $\frac{6}{4} = \frac{3}{?} =$ _____

25. $\frac{170}{51} = \frac{10}{?} =$ _____

28. $\dfrac{4}{45} = \dfrac{12}{?} =$ _____

26. $\dfrac{6}{39} = \dfrac{2}{?} =$ _____

29. $\dfrac{3}{?} = \dfrac{21}{28} =$ _____

27. $\dfrac{?}{39} = \dfrac{2}{13} =$ _____

30. $\dfrac{2}{13} = \dfrac{?}{26} =$ _____

3.4 Ratio Tables and Graphs

3.4.1 Ratio Tables

1. Eve reads 16 pages in 120 minutes. At this rate, how long will it take her to read 2 pages? _____.

2. To make a cup of orange juice, Tammy needs 2 spoons of sugar. At this rate, how many cups of orange would she make if she has 24 spoons of sugar? _____.

3. Use the ratio table to find how many tickets you need to sale to make a profit of $36? _____.

Number of tickets sold	1	3	—
Profit	6	18	36

4. Use the ratio table to find the missed value _____.

Time elapsed	1	2	5
Distance traveled	4	8	--

5. If your daily consumption of sugar is 8 ounces, how many days would 2 pounds of sugar lasts in this rate?_____

3.4.2 Graphing Ratio tables

Mark makes $8 an hour working at a restaurant. Bob makes $8.5 an hour working at another restaurant.

1. Make a table for each that shows the amount earned working 1, 2, and three hours_____.

2. How much will Mark make if he works for 8.5 hours? _____

3. How much will Bob make if he works for 9 hours? _____

4. After how long will Bob makes $102? _____

5. Will both earn the same amount at the same time? _____

6. List the ordered pairs for each relating hours and earnings_____.

7. Graph the ordered pairs for each on one coordinate plane (optional)_____

3.5. Percent to Fraction

3.5.1 Convert percent to fraction

1. 25% = _____   2. 50% = _____

3. 12% = _____

4. 68% = _____

5. 75% = _____
6. 86% = _____

7. 1 % = _____

8. 100 % = _____

9. 12.5 % = _____

10. 4.2% = _____

3.5.2 Convert the fraction to percent

1. 1/20 = _____

2. 30/100 = _____

3. 3/6 = _____

4. 25/100 = _____

5. 1/4 = _____

6. 3/4 = _____

7. 1/10 = _____

8. 12/48 = _____

9. 6/9 = _____

10. 2/5 = _____

11. 9/10 = _____

12. 11/10 = _____

13. 25/40 = _____

14. 2/25 = _____

3.6. Percent to Decimal

3.6.1 Convert percent to decimal

1. 43% = _____

2. 23% = _____

3. 22.1% = _____

4. 100% = _____

5. 2.01% = _____

6. 75% = _____

7. 541% = _____

8. 0.01% = _____

9. 101% = _____

10. 1% = _____

11. 6.5 % = _____

12. 99.9 % = _____

13. 10.1 % = _____

14. 0.28% = _____

3.6.2 Decimal to percent

Convert decimal to percent

1. 0.85 = _____

2. 0.01 = _____

3. 0.1 = _____

4. 1.23 = _____

5. 65 = _____

6. 4.67 = _____

7. 21.23 = _____

8. 0.99 = _____

9. 9.9 = _____

10. 99.9 = _____

11. 0.021 = _____

12. 2.08 = _____

3.7. Relate Fractions, Decimals, and Percent

3.7.1. Convert fraction to decimal then to percent

1. 3/4

   Decimal = _____

   Percent = _____

2. 3/8

   Decimal = _____

   Percent = _____

3. 12/48

   Decimal = _____

   Percent = _____

4. 3/24

   Decimal = _____

   Percent = _____

5. 11/110

   Decimal = _____

   Percent = _____

6. 1/8

   Decimal = _____

Percent = _____

7. 25/225

    Decimal = _____

    Percent = _____

8. 15/150

    Decimal = _____

    Percent = _____

9. 1/100

    Decimal = _____

    Percent = _____

10. 20/2000

    Decimal = _____

    Percent = _____

3.7.2. Convert fraction to percent then to decimal

1. 1/4

    Percent = _____ Decimal = _____

2. 35/100

    Percent = _____ Decimal = _____

3. 5/6

    Percent = _____ Decimal = _____

4. 1/50

    Percent = _____ Decimal = _____

5. 5/4

    Percent = _____ Decimal = _____

6. 11/22

    Percent = _____ Decimal = _____

7. 1/20

    Percent = _____ Decimal = _____

8. 5/40

    Percent = _____ Decimal = _____

9. 3/9

    Percent = _____ Decimal = _____

10. 2/5

    Percent = _____ Decimal = _____

3.8. Finding Percent of a Number

1. What is 53 % of 100? = _____

2. What is 12 % of 300? = _____

3. 10 is what percent of 200? = _____

4. 10 is _____ percent of 80.

5. What is 45 % of 90? = _____

6. 58% of 400 = _____

7. 28 % of 810 = _____

8. 27 % of 900 = _____

9. 175 is 35 % of _____

10. 99 is 30 % of _____

3.9. Comparing and ordering percent, fractions, and decimals

Use <, >, or = to make each true

1. 2/5 ____ 40%

2. 0.25 ____ 2/4

3. 0.650 ____ 0.675

4. 2/3 ____ 0.6

5. 0.7 ____ 8/10

6. 52% ____ 0.55

7. 80% ____ 8/10

8. 5.4 _____ 540%

9. 7/12 _____ 7/9

10. 1/6 _____ 6/31

Order the numbers from least to greatest if the numbers are different.

11. 1/7, 4/32, 0.56, 20% = _____

12. 3%, 0.07, 1/25 = _____

13. 3/9, 0.4, 25% = _____

14. 1.25, 89%, 110% = _____

15. 1/3, 2/3, ¾ = _____

16. 3/5, 37.5%, 3/7, 0.5 = _____

17. 9%, 0.9, 100%, 11/10 = _____

18. 26%, 0.31, 4/12, 0.6 = _____

19. 0.6, ¾, 6.7, 10/15 = _____

20. 3/5. 0.6, 60% = _____

3.10. Estimating with Percent

Estimate the percent of a number

1. Estimate 51% of 593 = _____
2. Estimate 40% of 39 = _____
3. Estimate 47% of 84 = _____
4. Estimate 19% of 41 = _____
5. Estimate 68% of 61 = _____
6. Estimate 18 % of 297 = _____
7. Estimate 19% of 52 = _____
8. Estimate 21% of 98 = _____
9. Estimate 37% of 119 = _____
10. Estimate 18% of 71 = _____

11. Estimate 69% of 298 = _____
12. Estimate 59% of 209 = _____
13. Estimate 24% of 298 = _____
14. Estimate 39% of 148 = _____
15. Estimate 29% of 201 = _____
16. Estimate 89% 902 = _____
17. Estimate 29% of 998 = _____
18. Estimate 61% of 303 = _____
19. Estimate 68% of 202 = _____
20. Estimate 9% of 0.001 = _____

3.11. Percent Word Problems

1. Samantha wants to spend 23% of her money for summer vacation. She has $4,600. How much money will she spend?

   Answer = _____

2. A T-shirt on sale costs 40% off the original price. If you pay $19.99 for the T-shirt. How much was the original price of the T – shirt?

   Answer = _____

3. Ms. Johnson wants to pay 35% of her mortgage. The current mortgage is 298,500. Estimate the amount she planned to pay

   Answer = _____

4. A bucket can hold 397 marbles. 59% of the marbles are green. About how many marbles in the bucket are green?

   Answer = _____

5. Trevon compared 20% of 40 and 30% of 25. He concluded 30% of 25 is greater. Is he correct?

   Answer = _____

6. In a test the grade of a student is 85 %. Each question is worth 5%. How many problems did the student miss?

   Answer = _____

7. There are 180 patients in a hospital for COVID – 19 tests. 150 were tested negative. What percent of the patients are free from virus? (Round to the nearest whole number)

8. 45% of a two-liter bottle is filled with organic olive oil. Determine the amount of oil in the bottle.

   Answer = _____

9. Chris planned to spend 20 % of his birthday gift buying a new game. He is thinking he will get $100 from his dad. $50 from his mom, and $30 from his sister. How much will he have after buying the new game? (Round to the nearest dollar)

Answer = _____

10. Chris changed his mind and wanted to save 95% of his birthday gift. He is thinking he will get $120 from his dad. $50 from his mom, and $30 from his sister. How much money will he save if he receives all the money he expected?

Answer = _____

11. A student's attendance was 95% of the 200 school days. How many days did the student attend school in the academic year?

Answer = _____

12. There is a huge sale at the mall. You can buy a phone case for $6 which is 20% of the original price. What was the original price of the phone case?

Answer = _____

13. Of ten friends, five of them like watching basketball games, one likes watching movies, and the others like playing cards. Find the percentage of basketball fans.

Answer = _____

14. There are 300 pens and pencils in a bag, of which 28% are pencils and 72% are pens. How many pens are in the bag?

Answer = _____

15. How much federal tax will you pay if your make $40,000 a year and the federal tax is 12%?

Answer = _____

16. If you drink 15% of gallon a day. How much water do you drink in 64 days?

Answer = _____

17. There are 84 children in a park. 21 of those students are girls. What percent of the them are boys?

Answer = _____

18. A soccer player made 3 out of 5 penalties. What percent of the penalties did the player miss?

Answer = _____

19. Paul bought a laptop that sells for $299.99. He has a. 21% discount coupon. About how much will he save?

Answer = _____

20. In question 19, about how much money did Paul spend for the laptop?

Answer = _____

3.12. Ratio & Proportion Problems

Use the table below to answer the questions

	Boys	Girls
2nd Grade	20	25
3rd Grade	30	35

1. Write each ratio as a fraction in simplest form.

   a. 2nd-grade boys to 3rd-grade boys _____

   b. 3rd-grade girls to 2nd-grade boys _____

   c. 2nd graders to 3rd graders _____

d. 3rd graders to 2nd graders _____

e. boys to girls _____

f. girls to all students _____

g. boys to all students _____

2. A 64-ounce container of sports juice costs $6.50. A 48-ounce container of the same juice costs $4.25. Which size container is cheaper to buy?

3. Solve each proportion.

a. $5/4 = r/8$

r = _____

b. $1/3 = 6/x$

x = _____

c. 2/5 = c/5

c = _____

d. 7/3 = 21/y

y = _____

e. 5/x = 25/25

x = _____

f. 49/x = 7/2

x = _____

g. 10/x = 2/3

x = _____

3. You can peel 8 potatoes in 27 minutes. How long will it take you to peel 16 potatoes?

Answer = _____

4. You can read 45 pages of a book in 2 hours. How many pages can you read in 6 hours?

   Answer = _____

5. Nine out of ten students prefer math class over lunch. How many students do not prefer math if 200 students were asked?

   Answer = _____

6. Two candies cost 15 cents. How many Candies can you buy if you have 90 cents?

   Answer = _____

7. In question 7, how much do you need to buy 100 candies?

   Answer = _____

8. If you rent eight DVDs for $ 5, how much do you pay to rent 32 DVDs?

Answer = _____

9. A phone company charges $4.5 for every 382 text messages. How much will you pay to text 15,471 messages?

Answer = _____

10. Two rectangular yards are proportional. The first yard is 20 meters by 8 meters, the second is 25 meters by x meters. Find the missed side, x

Answer = _____

11. Jonas bought three games for $ 222.8. How much will he pay if he wants to buy six games for his friends?

Answer = _____

12. If 5% of 100,000 people are unemployed, how many people are employed?

Answer = _____

3.13. Chapter 3: Test

1. If you run 2 miles per hour, how long will it take you to run 26 miles?

    a. 10 hours

    b. 5 hours

    c. 13 hours

    d. 6.5 hours

2. A classroom has 28 chairs, 8 tables and three whiteboards. What is the ratio of tables to chairs?

    a. 4/5

    b. 7/2

    c. 28/3

    d. 2/7

3. In question 2, what is the ratio of whiteboards to tables?

    a. 3/28

    b. 3/8

    c. 8/3

    d. 8/28

4. Tom runs 45 miles in five hour. What is his average speed?

    a. 5 mph (miles per hour)

    b. 9 mph

    c. 10 mph

    d. 15 mph

5. In question 4, how far will Tom go if he can run for 10 hours?

    a. 65 miles

b. 120 miles

c. 90 miles

d. 75 miles

6. Select ratio equivalent to 1/3

   a. 2/5

   b. 2/6

   c. 4/6

   d. 5/20

7. Which is not equivalent to 3/8?

   a. 24/9

   b. 6/16

   c. 9/24

   d. 12/32

8. Jenny needs 10 spoons of sugar to make 2 liters of juice. How many spoons of sugar she needs to make 200 liters of juice?

   a. 20

   b. 50

   c. 100

   d. 1000

9. Ben squeezed one lemon in a bottle of water. Sam squeezed 3 lemons in 5 bottles of water. Whose water tastes sourer?

a. Sam's

b. Ben's

c. Sam's water tastes the same as Ben's

d. No answer

10. The number 0.387 is in _____ form

   a. Fraction

   b. Decimal

   c. Equivalent

   d. Percent

11. Convert 30/50 to decimal form

   a. 60%

   b. 0.8

   c. 0.6

   d. 3/5

12. The chance to pass a test is 80%. What fraction is it?

   a. ¾

   b. 4/5

   c. 3/7

   d. 8/100

13. In order to get a grade of 'A' in math, you need to score 19.2 out of 20. What is your percentage score?

   a. 90%

   b. 97%

c. 87%

d. 96%

14. Bill studied 4/6 of the whole day for his final. How long did he study?

   a. 10 hours

   b. 16 hours

   c. 12 hours

   d. 15 hours

15. 35% of students in a school are boys. How many students are girls if the school has 300 students?

   a. 195

   b. 200

   c. 105

   d. 70

16. Enter the correct sign if 0.38 _____ 5/11.

   a. <

   b. =

   c. >

17. Which of the following is equivalent to 3/4 in decimal and percent form?

   a. 0.75 and 75%

   b. 0.65 and 65%

   c. 0.75 and 70%

   d. 0.85 and 75%

18. What percent of 700 is 196?

   a. 25%

   b. 26%

   c. 27%

   d. 28%

19. 65% of _____ is 325.

   a. 200

b. 500

c. 600

d. 700

20. Which is the correct equivalent expression?

    a. 0.2, ½, 20%

    b. 0.1, 20/ 200, 10%

    c. 0.2, 1/5, 10%

    d. ¾, 75%, 0.6

21. Which is the correct increasing order?

    a. 4.5, 30%. 5/6, 20%

    b. 0.45, 60%, ¾, 80%

    c. 2.3, 3.2, 28%, 1/5

    d. 3%, 0.03, 5, 2.6

22. Which is the percent form for 2.18?

    a. 21.8%

    b. 2.18%

    c. 218%

    d. 23%

23. Four out five students passed the Math test. What percentage of the students failed?

    a. 1/5

    b. 80%

    c. 20%

    d. 75%

24. Solve the proportion to find x, $\dfrac{5}{30} = \dfrac{6}{x}$

   a. x = 10

   b. x = 36

   c. x = 90

   d. x = 60

25. If you are allowed to take 20 % of 10% of the price of a car as a commission. How much will you get if you sell a car for $ 60,000?

   a. 6,000

   b. $1,200

   c. $3,000

   d. $1 800

CHAPTER 4

EXPRESSIONS & EQUATIONS

4.0 Review Notes on Expressions and Equations

- Numbers like: 3^5, 10^3 are in exponent forms. 5^3 means 5x5x5 = 125. 5^3 is read as 5 to the exponent 3 or 5 the power of 3.

- **A numerical expression**: a combination of numbers and operations. Use order of operations to evaluate numerical expressions.
- **Variable:** unknown number, usually represented by letters like x, y, z.
- **Algebraic expression:** combination of numbers, variables and operations like (+, -, ×, ÷)
- **Equation:** a mathematical statement that includes an equal sign. Ex. 7x + 3 = 11
- **Solving** an equation means finding the number that makes the equation true.
- **Evaluating an expression:** Finding the value of the given expression by replacing number for the variable
- **Simplifying expression:** writing the expression in simple form without changing it. Ex. (4x/2) can be written in a simplified form as 2x.
- **Addition Principle**: Adding the same number on both sides of an equation does not change the equation. Ex. Given x + 1 = 4. If you add 3 on both sides of the equation, you get x +4 = 7. The equality sign is not affected.
- **Subtraction Principle**: Subtracting the same number on both sides of an equation does not change the equation. Ex. Given 2x = 5. If you subtract 3 on both sides of the equation, you get 2x -3 = 2. The equality sign is not affected.
- **Multiplication Property**: Multiplying each term of an equation by the same number does not change the equation. Ex. Given 3x + 2 = 5. Multiplying each term by 4 we get 12x + 8 = 20. The equality sign is not affected.
- **Division Property**: dividing each term of an equation by a nonzero number does not change the equation. Ex. Given 4x + 2 = 6. Dividing each term by 2 we get 2x + 1 = 3. The equality sign is not affected.
- **Commutative property of addition**:

    a + b = b + a. Ex. 3 + 4 = 4 +3
- **Commutative property of multiplication**:

    a × b = b × a
- **Associative property addition**:

$(a + b) + c = a + (b + c)$

- **Associative property of multiplication:**

  $(a \times b) \times c = a \times (b \times c)$

- **Identity property of addition:**

  $a + 0 = a$

- **Identity property of multiplication:**

  $a \times 1 = a$

- **Distributive property of multiplication over addition:**

  $a \times (b + c) = (a \times b) + (a \times c)$

- **Factoring** is writing the expression as a product of two or more terms by separating the common number or variable in the expression.

4.1. Exponent and product

Write each product as an exponent

1. $3 \times 3 \times 3 =$ _____

2. $5 \times 5 \times 5 \times 5 \times 5 =$ _____

3. $0.4 \times 0.4 \times 0.4 \times 0.4 =$ _____

4. $(1/2) \times (1/2) \times (1/2) =$ _____

Find the value of each product

5. $8 \times 8 \times 8 =$ _____

6. $2 \times 2 \times 2 \times 2 \times 2 = $ _____

7. $0.3 \times 0.3 \times 0.3 = $ _____

8. $(1/2) \times (1/2) \times (1/2) = $ _____

Write each power as a product

9. $4^2 = $ _____

10. $(0.1)^3 = $ _____

11. $1^5 = $ _____

12. $(1/4)^3 = $ _____

Find the value of the exponent

13. $3^0 = $ _____

14. $1^{10} = $ _____

15. $2^5 = $ _____

16. $10^3 = $ _____

17. $0^6 = $ _____

18. $(1/2)^3 = $ _____

19. $(2.3)^2 = $ _____

20. $(1.01)^2 = $ _____

4.2. Numerical Expressions

Find the value of each numerical expression.

1. $5 + 4 - 3 = $ _____

2. $(23 - 10) + 2 \times 4 = $ _____

3. $4 + 8 \times (5 + 4) = $ _____

4. $24 - 2^4 \div 8 = $ _____

5. $4^3 - 18 \div 3 = $ _____

6. $5 \times (3^2 - 8) \div 2 = $ _____

7. $6 + 5^2 \div (9 - 4) \div 5 + 3 = $ _____

8. $44 \div 4 \times (1 \div 11) = $ _____

9. $27 \div (4 + 5) \times 5 - 11 = $ _____

10. $2^2 + 3^3 + 4^4 = $ _____

11. $20 + 18 \div 2 \times 9 - 5 = $ _____

12. $5 \times 14 + 12 \div 3 = $ _____

13. $18 \times 4 + 18 = $ _____

14. $10 \times 12 - 9 \times 12 = $ _____

15. $12 \times 20 + 7 \times 5 = $ _____

16. $6 \times 7 \times 15 + 218 = $ _____

17. $200 - 24 \div 8 \times 6 = $ _____

18. $8 \times 1 \times 14 + 18 \div 2 = $ _____

19. $20 \times 12 - 14 \div 2 + 15 = $ _____

20. $18 \div 2 - 3 + 3 = $ _____

21. $39 + 15 \div 5 \times 2 = $ _____

4.3. Variables, Expressions and Equations

4.3.1. Identify the variable in the expression, if any.

1. $10x + 83 =$ _____

2. $12a + 4x - 300 =$ _____

3. $5000 + 25 =$ _____

4. $4a + 3b + 6c - 100 =$ _____

5. $2x - 3y =$ _____

6. $780y - 67x + 3p =$ _____

4.3.2. Identify if each of the following is an equation or expression

1. $8 + x = 9 =$ _____

2. $9x + 2y =$ _____

3. $10a - 3b - 6c =$ _____

4. $5x - 2y = 3x =$ _____

5. $4x - 2z + 3y =$ _____

4.3.3 List the operations used in each expression

1. 72x – 6y = _____

2. 2a + 4x – 300 = _____

3. 43x ÷ 25 = _____

4. 4a + 3b + 6c = _____

5. 2 ÷ 3z = _____

6. 4x – 3 = _____

7. 8x – 2z ÷ 3y = _____

8. -9x + 12y = _____

9. 3x – 2z ÷ 4 = _____

10. 3x + y ÷ 7z = _____

11. -3x ÷ 2y – 4 = _____

12. 11x + 2y + 3z – 3a = _____

4.4. Evaluating Expressions.

Evaluate each expression. Use x =1, y =2, z = -1, a = -2, b = 3, d = ½, and c = 0

1. $10x + 9 =$ _____

2. $7x^2 - 2d =$ _____

3. $8z + 3y^3 - b^2 =$ _____

4. $xyz - xz + 6d =$ _____

5. $2xy + 3yd - 3y =$ _____

6. $xy \div yz =$ _____

7. $2x + 3 =$ _____

8. $7a + 4x - 300 =$ _____

9. $5000 + 25 =$ _____

10. $4a + 3b + 6c - 100 =$ _____

11. $2x - 3y =$ _____

12. $3x + 9 =$ _____

13. $4x + 2y - b =$ _____

14. $8a - 3b - 6c =$ _____

15. $7x - 2y - z =$ _____

16. $5x - 2z + 3y =$ _____

17. $2x - 6y =$ _____

18. $2a - 4x - 300 =$ _____

19. $43x \div 25 =$ _____

20. $4a + 3b + 6c =$ _____

21. $2 \div 3z =$ _____

22. $4x - 3 =$ _____

23. $8x - 2z \div 3y =$ _____

24. $-9x + 12y =$ _____

25. $3x - 2z \div 4 =$ _____

26. $3x + y \div 7z =$ _____

27. $-3x \div 2y - 4 =$ _____

28. 11x + 2y + 3z – 3a = __   30. 5x – 3z + 5c = _____

29. – 6x – 3z + $\frac{y}{x}$ = _____

4.5. Writing Algebraic Expression

1. Seven times a number plus eight.

2. Five less than a number.

3. Three less than one fourth of number is 8.

4. Four times a number.

5. Six more than twice as many as a number

6. A number plus five = _____

7. Twice of a number is ten = _____

8. The difference between two numbers is five = _____

9. Five minus a number is 20 = _____

10. A number is added to 3 to get 8 = _____

11. 10 is taken away from two times a number to get 5 = _____

12. A number doubled is 20 = _____

13. The sum of two consecutive numbers is 25 = _____

14. Ben drinks 3 liters of water every day for two weeks = _____.

15. Jenny is thinking of a number. 1/6 of the number is 5 = _____.

4.6. Properties

Name the property used in each equation.

1. $11 \times 43 = 43 \times 11$ _____

2. $5 + 6 = 6 + 5$ _____

3. $x + y = y + x$ _____

4. $a \times b = b \times a$ _____

5. $(2 + 3) + 5 = 2 + (3 + 5)$ _____

6. $(a + b) + c = a + (b + c)$ _____

7. $5 \times (3 + 4) = (5 \times 3) + (5 \times 4)$ _____

8. $z \times (y + x) = (z \times y) + (z \times x)$ _____

9. 3 × (6 + 7) = (3 × 6) + (3 × 7) _____

10. (5 × 4) × 8 = 5 × (4 × 8) _____

11. 23 × 9 = 9 × 23 _____

12. (a × b) × c = a × (b × c) _____

4.7 Equivalent Expressions

Write an equivalent expression to each that does not have parentheses

1. 4 + (y + 8) = _____

2. (3 × y) × 8 = _____

3. (5 − x) + 2 = _____

4. 2 × (x + y) = _____

5. (y ÷ 1) × 3 = _____

6. (3 × y) × 0 = _____

7. 25 − (x − 4) = _____

8. (4 − 2 − x) + 4 = _____

9. 4 (5 + 6) = _____

10. 2 (x − 2) = _____

11. 3(14 + 6) = _____

12. 5 (x + y) = _____

Determine whether the two expressions are equivalent. If so, tell what property is applied.

13. 25 + 0 and 25 _____

14. 45 ÷ 9 and 9 ÷ 45 _____

15. (6 + 5) + 4 and 6 + (5 + 4) _____

16. (19 − 13) − 5 and 19 − (13 − 3) _____

17. 4 × (5 × 6) and (4 × 5) × 6 _____

18. (30 ÷ 5) ÷ 6 and 30 ÷ (5 ÷ 6) _____

19. 23 × 1 and 23 _____

20. 125 ÷ 5 and 5 ÷ 125 _____

4.8. Factoring Expression

4.8.1. Use distributive property to rewrite each algebraic expression given below

1. $5(2 + 4) =$ _____

2. $3(x + 2) =$ _____

3. $2(x + 4) =$ _____

4. $6(5 + 4y) =$ _____

5. $4(3x + 5) =$ _____

6. $8(2x + 3y) =$ _____

7. $3(16x + 2y) =$ _____

8. $0.5(8x + 10y) =$ _____

9. $(1/4)(32 + 16x) =$ _____

10. $0.1(10x + 20y) =$ _____

11. $2(2x + 3y + 4w) =$ _____

12. $3(3x + 3y + 4z) =$ _____

13. $0.25(4x + 8y + 12z) =$ _____

14. $2(3 + 5x + 2y) =$ _____

15. $3(2x + 3 + 5y) =$ _____

4.8.2. Factor each expression

1. 15 + 21 = _____

2. 7 + 49 = _____

3. 16 + 24 = _____

4. 4x + 16 = _____

5. 5x + 10 = _____

6. 60 + 24x = _____

7. 56x + 49 = _____

8. 6x + 9y = _____

9. 70x + 25y = _____

10. 8x + 36y = _____

11. 4x + 6y + 14z = _____

12. 3x + 6y = _____

13. 20x + 18y + 36 = _____

14. 2x + 4y + 8 = _____

15. 5x + 25y + 75 = _____

4.9. Simplifying expressions with one and two variables

4.9.1 Find the simplified number.

1. $4 \times 5 =$ _____

2. $(3 \times 6) \times 4 =$ _____

3. $(89 + 34) + 6 =$ _____

4. $(2 \times 5) + (2 \times 7) =$ _____

5. $6 \times (4 + 3) =$ _____

6. $14 + (12 + 13) =$ _____

7. $(5 + 9) \times 4 =$ _____

8. $(4 \times 5) + 2 \times (3 + 5) =$ _____

9. $(2 \times 5) + (5 \times 2) =$ _____

10. $(7 + 2) \times (4 + 3) =$ _____

4.9.2 Use one of the symbols <, >, or = to compare the numbers

1. 5×6 _____ 6×5

2. (5 + 4) × 2 _____ 5 + (4 × 2)

3. (2 + 3) + 4 _____ 2 × (3 + 4)

4. 6 × (3 + 7) _____ (6 × 3) + (6 × 7)

5. (3 × 9) ×7 _____ 9 × (3 × 7)

6. 11 × (8 + 7) _____ 11 ×8 + 11 × 7

7. 6 × (8 + 7) _____ 6 + 8 + 6 × 7

8. 9 + (6 + 2) _____ 9 + 6 x 2

9. 9 × 10 _____ 10 + 9

10. (7 × 10) + (6 × 10) _____ 13 × 10

4.9.3. Simplify each expression with one variable

1. 5(4x) = _____

2. 2 + 4x +6 = _____

3. 3x + x + 5x = _____

4. 12x + 13 + 4x = _____

5. 3(x + 4) + 8 = _____

6. 4(3x + 2) + 3x = _____

7. 3(3x +2) + 4(2x) = _____

8. 1(3x + 4) + 4 = _____

9. 0(55x + 99) + 7x = _____

10. 5(4 + 2x) + 3(7 + 3x) = _____

4.9.4 Simplify each expression with more variables.

1. 3(x + y) +2y = _____

2. 5(3x) + 2y + 5x = _____

3. 4 (2+ 3x) + 5x = _____

4. (2y + x) + 24y = _____

5. 2(4x + 2y) + 10x = _____

6. 4x + 2y + 7x = _____

7. 3(2x + 9y) = _____

8. 3(2x) + 4(8y) = _____

9. 2(3x + 4) + 4(2x + 3y) = _____

10. 5x + 3y + 13x + 5y = _____

4.10. Solving Equations: Addition Principle

4.10.1 Add the same number on both sides of the equation to solve each equation.

1. x – 1 = 6 _____

2. y – 5 = 11 _____

3. x – 13 = 3 _____

4. y – 25 = 5 _____

5. z – 26 = 14 _____

6. x – 29 = 21 _____

7. x – 10 = 90 _____

8. x – 90 = 10 _____

9. x − 28 = 72 _____

10. x − 34 = 166 _____

4.10.2. Subtract the same number on both sides of the equation to solve each equation.

1. x + 2 = 4 _____

2. y + 4 = 10 _____

3. x + 25 = 42 _____

4. y + 6 = 55 _____

5. z + 6 = 16 _____

6. x + 9 = 21 _____

7. x + 20 = 70 _____

8. x + 50 = 60 _____

9. x + 18 = 78 _____

10. x + 34 = 46 _____

4.11. Solving Equations: Multiplication Principle

4.11. Multiply by the reciprocal or divide by the same number each side of the equation to solve each equation.

1. 4x = 16 _____

2. 8y = 32 _____

3. 4x = 64 _____

4. 5y = 55 _____

5. 6z = 60 _____

6. 7x = 21 _____

7. 8x = 80 _____

8. 7x = 49 _____

9. 9x = 81 _____

10. 10x = 100 _____

11. -5x = 25 _____

12. -3y = -9 _____

13. -4x = 16 _____

14. -8x = -24 _____

15. -0.5 x = 4 _____

16. 3x = - 27 _____

17. – 12x = 144 _____

18. 13x = 169 _____

19. - 25x = 200 _____

20 - 30y = -300 _____

4.12. Multistep Equations

4.12.1 Check if the equation is true for the given value (s).

1. x + 3 = 99, x = 95

Answer = _____

2. 2x – 4 = 0, x = 2

Answer = _____

3. -x + 6 = 8, x = -3

Answer = _____

4. 3y = 81, y = 27

Answer = _____

5. 4y – 5 = 15, y = - 5

Answer = _____

6. -6x + 5 = 25, x = 5

Answer = _____

7. - 8x + 8 = 72, x = - 8

Answer = _____

8. 2y – 32 = -40, y = 4

Answer = _____

9. – 2y + 32 = 60, y = -14

Answer = _____

10. –x – 1 = -3, x = 2

Answer = _____

11. 2 – y = 7, y = 5

Answer = _____

12. 5 – y = 15, y = -10

Answer = _____

13. 2 + 7x = 30, x = 4

Answer = _____

14. 5 – 6x = 10, x = 1

Answer = _____

15. -6x + 6 = 42, x = -7

Answer = _____

16. 12 – 7x = 12, x = 1

Answer = _____

17. 19y – 16 = -16, y = 0

Answer = _____

18. 20x + 24 = 44, x = 1

Answer = _____

19. -4y – 16 = - 32, y = 4

Answer = _____

20. -0.5y + 3.5 = -1, y = 9/2

Answer = _____   Answer = _____

4.12.2 Solve the multistep equations

1. $2x + 4 = 10$ 9. $-2y + 32 = 60$

Answer = _____   Answer = _____

2. $2x - 4 = 0$ 10. $-2x - 1 = -3$

Answer = _____   Answer = _____

3. $-3x + 6 = 9$ 11. $3 - 2y = 7$

Answer = _____   Answer = _____

4. $3y - 4 = 8$ 12. $15 - 3y = 15$

Answer = _____   Answer = _____

5. $-4y - 5 = 15$ 13. $2 + 7x = 30$

Answer = _____   Answer = _____

6. $-6x + 5 = 35$ 14. $5 - 6x = 35$

Answer = _____   Answer = _____

7. $-8x - 8 = 72$ 15. $-6x + 6 = 42$

Answer = _____   Answer = _____

8. $2y - 32 = 0$ 16. $12 - 7x = 12$

Answer = _____   Answer = _____

17. 19y – 16 = 22

Answer = _____

18. 20x + 24 = 44

Answer = _____

19. -4y – 16 = - 32

Answer = _____

20. -0.5y = -1

4.13. Formulas

Use the given formula to find the value of the remaining variable

1. $A = \dfrac{b \times h}{2}$, b = 5, h = 8. Find A

2. P = 2l + 2w, P = 40, l = 10, Find w

3. A = l × w, l = 20, w = 6, Find A

4. D = s/t, t = 4, s = 100, Find D

5. A = 2x + y − 4z, A = 30, x = 5, y = 0, Find Z

6. M = DV, M = 30, D = 5, Find V

7. C = 6.28 × r, r = 3, Find C

8. V = l × w × h, v = 72, l = 2, w = 6, Find h

4.14. Expression Word Problems, Use variable for unknown number.

1. Ten friends paid x dollars each for admission to a talent show. The total amount paid was $120. Write the equation defining it and find how much each person paid for the show?

    Answer = _____

2. Jackson is x years old. His sister is 4 years older than Jackson. The sum of their ages is 50. How old is Jackson's sister?

    Answer = _____

3. The sum of a number and 13 is 24. What is the number?

Answer = _____

4. Three times a number is 30. What is the number?

Answer = _____

5. The difference between a number and 3 is nine. What is one possible number?

Answer = _____

6. Ten plus a number is six. What is the number?

Answer = _____

7. A number is multiplied by 7 is 35. What is the number?

Answer = _____

8. A positive number minus three is 11. What is the number?

Answer = _____

9. Nancy has $6 dollars. Her older brother gave her some money. Now she has $ 17. How much did Nancy get from her brother?

Answer = _____

10. The sum of two consecutive numbers is 13. What are the two numbers?

Answer = _____

11. Sam reads 102 pages a day. How many pages can he read in four days?

Answer = _____

12. Chris is thinking of a number. Twice of Chris's number plus 3 is 15. What number is Chris thinking of?

Answer = _____

4.15. Chapter 4 Test

1. Factor: $3x + 6$.

    a. $3(x + 6)$

    b. $3(2x + 1)$

    c. $3(x + 2)$

    d. $2(x + 3)$

2. Which of the following is representing an algebraic expression?

    a. $3y + 1 = 7$

    b. $5x - 3 < 24$

    c. $y - 1 = 2x$

    d. $3y + 2$

3. Which of the following is representing an equation?

    a. $3x + 5$

    b. $9x + 6 = 7$

    c. 33

    d. $y < 5$

4. Name the property used in the equation: $10 \times 5 = 5 \times 10$.

    a. Commutative property of addition

    b. Commutative property of multiplication

    c. Associative property of multiplication

d. Distributive property of addition

5. Name the property used in the equation: $(3 + 4) + 5 = 3 + (4 + 5)$.

   a. Commutative property of addition

   b. Commutative property of multiplication

   c. Associative property of addition

   d. Distributive property of addition

6. Evaluate the expression: $3x - 10$ for $x = 6$.

   a. 18

   b. 8

   c. 6

   d. 28

7. Evaluate the expression: $2x - 10y + 6$, for $x = 2$ and $y = -1$.

   a. 20

   b. 0

   c. 12

   d. 4

8. Another form of writing: $(2 \times 5) + (2 \times 8)$ is _____.

   a. $2 \times (5 \times 8)$

   b. $2 \times (5 + 8)$

   c. $(2 \times 5) + 8$

   d. $2 \times (5 \times 8) \times 2$

9. Solve each equation: $x - 4 = 11$

   a. 10

b. -7

c. 15

d. 10

10. Solve each equation: x + 3 = -2

    a. - 3

    b. 2

    c. 9

    d. -5

11. Use one of the symbols <, >, or = to compare the numbers

(8 ×5) × 7 _____ (16 ×10) + (4 × 35)

    a. =

    b. >

    c. <

    d. None of the above.

12. The expression: x – (- 5) is the same as:

    a. x -5

    b. –x + 5

    c. x + 5

    d. – x - 5

13. Solve the equation: 5x = 35

    a. 5

    b. 6

    c. 8

    d. 7

14. The equation is true 2x − 6 = 12, for x = ___.

    a. 10

    b. 8

    c. 9

    d. − 9

15. 5x + 1 = 1 for x = _____

    a. 3

    b. 0

    c. -5

    d. -2

16. -3x + 9 = 12 for x = _____

    a. -2

    b. 1

    c. 0

    d. -1

17. Use the formula: P = 3w + h to find h when P = 20 and w = 5

    a. h = 4

    b. h = 9

    c. h = 5

    d. None of the above

18. Write expression describing the sentence. A number added to 8.

    a. 8x

    b. -8 + x

    c. 8 + x

    d. 8x + 8

19. Write the equation describing the sentence: the product of a number and three is 7.

    a. 7x = 3

    b. 3x = 7

    c. x = 3 × 7

    d. 3x = 8

20. Sam and Jill have $20. If Sam has two more dollars, how much does Sam have?

    a. 10

    b. 12

    c. 8

    d. 11

21. Twice of a number is added to 8 to get 38. What is the number?

    a. 15

    b. 20

    c. 30

    d. 12

CHAPTER 5: FUNCTIONS AND INEQUALITIES

5.0 Review Notes on Functions and Inequalities

- A relation is a set of order pairs.

- A function is a relation that assigns exactly one output value to one input value.

- A function rule is a rule relating the input to the output
- A function table is an organized way of displaying input and output values in a chart.
- Independent variable in a function is the input value
- Dependent variable in a function is the output value that depends on the input value.
- A sequence is a list of numbers in a specific pattern, so a sequence is a function.
- An arithmetic sequence is a sequence of numbers ordered in a way that the next number is found by adding the same number to the previous number or term.
- A geometric sequence is a sequence where the next term is found by multiplying the previous term by the same number.
- If the graph of a function is a line, the function is called a linear function.
- The four inequality signs are: < less than, > greater than, ≤ less than or equals, and ≥ greater than or equals.
- An inequality is a mathematical expression that has inequality sign.

5.1 Definition of a function

Write true or false.

1. A function is a relation that assigns exactly two output value to one input value _____

2. A function rule describes the relation between dependent and independent variable _____

3. In a function, the output is the dependent variable _____

4. In a function, the input is the dependent variable _____

5. If the output is defined by $3x - 1$ where x is the input variable. The output is 4 when the input 3 _____.

6. In question 5, the input is 6 when the output is 17_____

7. Given the function y = 3x + 3, if the input, x = 5, the output, y is 19_____

8. If the function is represented by the set of points (1, 2), (2, 4), and (3, 6). Input values are 1, 2, and 3_____.

9. In question 8, output values are 2, 4, and 6.

10. In question 8, the rule relating the input to the output is defined by y = 2x_____

5.2 Function tables and rules.

5.2.1. Fill the output based on the given function rule.

1.

Input (x)	Output (x +4)
1	1+4 =5
2	
3	
4	
5	

2.

Input(x)	Output 4x
2	8
4	
6	
8	

10	

3.

Input(x)	Output 25x +3
1	
2	
3	
4	
5	

4.

Input(x)	Output -2x -3
-3	
-2	
0	
1	
2	

5. Find the output, x + 4 if the input is x = 2 _____

6. Find the output, 3x - 4 if the input is x = 0 _____

7. Find the output, 2x + 5 if the input is x = 7 _____

8. Find the output, 10x - 4 if the input is x = 1 _____

9. Find the output, 8x - 7 if the input is x = 3 _____

10. Find the output, 11x, if the input is x = 4 _____

5.2.2 Find the rule for the function, output based on the given table values

1.

Input, x	1	2	3	4	5
Output	5	10	15	20	25

_____

2.

Input, x	1	2	3	4	5
Output	4	5	6	7	8

_____

3.

Input, x	1	2	3	4	5
Output	3	5	7	9	11

_____

4.

Input, x	1	2	3	4	5
Output	5	8	11	14	17

_____

5.

Input, x	1	2	3	4	5
Output	1	5	9	13	17

_____

5.3 Arithmetic and Geometric Sequences

Determine if each is an arithmetic or geometric sequence or neither.

1. 2, 4, 6, 8, 10, 12, 14 _____

2. 1, 4, 8, 12, 16, 20 _____

3. 2, 5, 8, 11, 14, 17 _____

4. 1, 1.1, 1.2, 1.3, 1.4, 1.5 _____

5. 1, 0.1, 0.01, 0.001, 0.0001, 0.00001 _____

6. 1, 10, 100, 1000, 10,000, 100,000 _____

7. 1, 2, 3, 4, 5, 6, 7, 8, 9 _____

8. 2, 4, 6, 7, 8, 10, 15 _____

9. 1, 1.5, 2, 2.5, 3, 3.5, 4 _____

10. 1, ½, ¼, 1/8, 1/16 _____

Determine the next term of each sequence.

11. 1, 7, 13, 19, 25, 31, ___

12. 3, 9, 27, 81, ___

13. 10, 9, 8, 7, 6, 5, ___

14. 2.5, 2.6, 2.7, 2.8, 2.9, ___

15. 1, 2, 3, 4, 5, 6, ___

16. 16, 8, 4, 2, 1, ___

17. 1, 5, 25, 125, ___

18. 0, 11, 22, 33, 44, 55, 66, ___

19. 3.25, 3.75, 4.25, 4.75, ___

20. 3, 15, 27, 39, ___

Find the missed term (number) in the sequence.

21. 45, 40, 35, 30, ____, 20, 15, 10

22. 2, 4, 8, 16, ____, 64, 128

23. 7, 14, ____, 28, 35, 42

24. 7, 8.2, 9.4, 10.6, ____, 13, 14.2, 15.4

25. 27, 9, 3, ____, 1/3, 1/9

26. 28.4, 25.6, ____, 20

27. 0.1, 0.2, 0.3, ____, 0.5

28. 1.5, 3, 4.5, ____, 7.5, 9

29. 384, 96, 24, ____, 1.5

30. 2401, ____, 49, 7, 1

5.4. Inequalities

Determine if the given number satisfies the inequality.

1. $x + 4 \leq 3$, $x = 2$ _____

2. $3x - 5 > 6$, $x = 4$ _____

3. $-2x + 3 < 12$, $x = -3$ _____

4. $4x - 7 \geq 24$, $x = 10$ _____

5. $3 - x < 3$, $x = -1$ _____

6. $2 + 5x > -3$, $x = -2$ _____

7. $3x + 9 \leq 30$, $x = 7$ _____

8. $4 - 2x > 2$, $x = 1$ _____

9. $2x + 2y < 5$, $x = 1$, $y = 2$ _____

10. $3y^2 - 5x > 0$, $y = 2$, $x = 1$ _____

State two numbers satisfying each inequality. Answer varies.

11. $X + 9 > 15$ _____

12. $2x - 8 < 0$ _____

13. $3x - 13 > 4$ _____

14. $-5 + x \geq 4$ _____

15. $2 - 4x < 10$ _____

16. $6y > 19$ _____

17. $3x - 5 > 23$ _____

18. $4x + 5 < 2$ _____

19. $9x + 33 > 51$ _____

20. $23x < 36$ _____

5.5. Writing Inequalities

Use x for the variable to write inequality for each question.

1. You must be at least 5 years old to go to elementary school_____.

2. The page contains at most 200 letters_____.

3. The game will take no more than 45 minutes_____.

4. My height is at least 6ft. _____.

5. Chris studied for less than two hours_____.

6. Jenny needs to do more than 10 questions to finish her homework_____.

7. Sally runs no more than 3 miles a day_____.

8. The number is greater than 6_____.

9. The number is not greater than 4_____.

10. A number is greater than 2 but not more than 7_____.

5.6 Graphing Inequalities

5.6.1 Write the opposite of each inequality

1. $x < 3$

2. $x > 2$

3. $x \leq 5$

4. $x \geq 4$

5. x < 0

6. x > - 2

7. x ≤ - 3

8. x – 1 > 3

9. x + 2 < 1

10. x – 2 ≥ 1

5.6.2. Write the inequality for each graph below.

1.

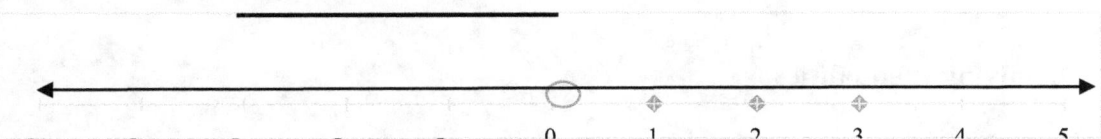

2.

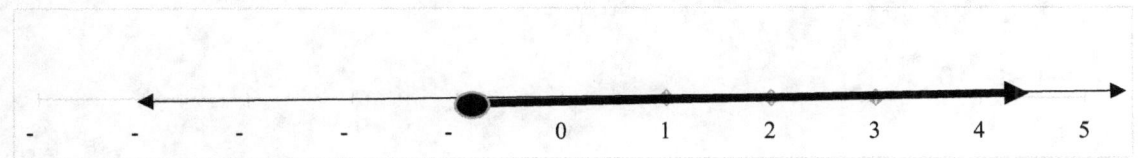

3.

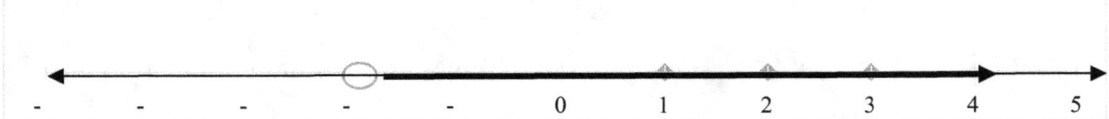

4.

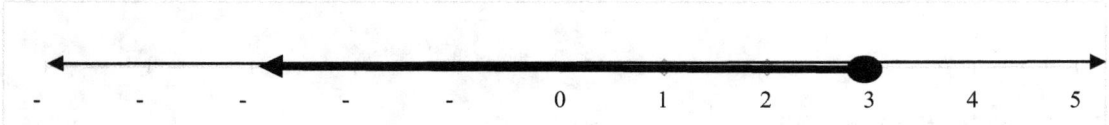

5.7. Solving Inequalities

1. $x - 1 < 5$ _____

2. $x + 1 > 3$ _____

3. $2x \leq 4$ _____

4. $5x \geq 10$ _____

5. $x + 4 < -10$ _____

6. $\dfrac{x}{2} < 1$ _____

7. $9x < 99$ _____

8. $\dfrac{2}{3}x < 6$ _____

9. $-x < 7$ _____

10. $-2x \geq 8$ _____

11. $-x - 1 < -15$ _____

12. $-x + 1 < 4$ _____

13. $\dfrac{-4}{6}x < 10$ _____

14. $-7x \geq 35$ _____

15. $-6x \leq -30$ _____

16. $x - 1 + 4x < 4$ _____

17. $3x - 1 - 2x < 5$ _____

18. $8x - 2x < 12$ _____

19. $2x - 1 < 7$ _____

20. $-3x + 1 \geq 5$ _____

5.8. Word Problem Involving Inequalities and Functions

1. Five added to the product of 2 and x is less than 11. Write the inequality and solve the problem_____.

2. A soccer ball costs $5. If Mark can spend at most $30. Write and solve the inequality to find how many soccer balls he can buy?

3. In order to pass grade 6 math exam, a student has to score 75 or more. Write an inequality to represent the situation.

4. An author receives a royalty of $15 for every book the publisher sells. The author wants to make at least $1500. Write and solve an inequality to find the number of books the publisher must sell.

5. The sum of three numbers, x, x + 1, and x + 2 is at least 102. Write and solve an inequality to find the minimum integer value of x satisfying the inequality.

6. Is the inequality x < 5 different from 5 > x?

7. A bag of chips sells for $3.5. How many bags of chips could you buy if you want to spend less than $311.5?

8. You want to buy movie tickets for each of your six friends. You have $120 to spend for the tickets. If a ticket costs $22.5, will you be able to pay for all your friends? Why?

9. Write a word problem using the inequality x – 3 < 7.

10. Write an inequality for the sentence: My weight plus 6 pound is more than 160 pounds.

5.9. Test

1. The output is defined by 3x – 1 where x is the input variable. If the input is 9, the output is_____

a. 30

b. 26

c. 29

d. 9

2. The function rule is y = 3x + 1, if y = 10, x = ____

   a. 31

   b. 9

   c. 3

   d. 4

3. The output of the function is:

   a. The independent variable

   b. The input variable

   c. The dependent variable

   d. Always positive

4. Find the output, 2x + 5, when the input, x = 7 is _____

   a. 17

   b. 18

   c. 16

   d. 19

5. The function rule representing the table below is:

Input, x	1	2	3	4	5
Output	2	5	8	11	14

a. x + 2

b. 3x - 1

c. 4x

d. 10x - 4

6. The sequence 0.2, 0.02, 0.002, 0.0002, 0.00002 is:

    a. Geometric sequence

    b. Arithmetic sequence

    c. Not geometric sequence

    d. Neither geometric nor arithmetic sequence

7. What is the next number in this pattern: 0, 11, 22, 33, 44, 55, 66, ____

    a. 72

    b. 75

    c. 77

    d. 79

8. What is the next number in this pattern: 3.25, 3.75, 4.25, 4.75, ____

    a. 4.95

    b. 5.05

    c. 5.25

    d. 5.75

9. 27, 9, 3, ____, 1/3, 1/9 is a geometric sequence. What is the missed number?

a. ¼

b. 1/3

c. 1

d. 3/2

10. Which number is NOT a solution of the inequality $5 - 2x \geq -3$?

   a. 0

   b. 4

   c. -5

   d. 6

11. Determine if x = - 2 satisfies the inequality -2x + 4 < 18.

   a. True

   b. False

12. Two numbers satisfying $3 - 2x > 5$.

   a. 1 and 2

   b. 3 and 4

   c. 0 and 1

   d. – 2 and - 3

13. Write the inequality representing: Five added to a number is at most 200

   a. X + 3 > 20

   b. X + 5 ≤ 200

   c. X – 5 > 200

    d. $2x + 2 < 200$

14. Solve the inequality: $X - 2 > 4$

    a. $X > 6$

    b. $X < 2$

    c. $X > 10$

    d. $X < 10$

15. Solve the inequality: $X + 3 < 5$

    a. $X < 2$

    b. $X < 3$

    c. $X < 4$

    d. $X < 1$

16. Solve the inequality: $2X - 3 > 7$

    a. $X < 4$

    b. $X > 4$

    c. $X > 5$

    d. $X < 5$

17. Solve the inequality: $-3X + 4 < 13$

    a. $X > -3$

    b. $X < -3$

    c. $X > 3$

    d. $X < 3$

18. The sum of a number and 32 is greater than the product of −3 and that number. What are the two possible values satisfying the inequality?

   a. -10 and −11

   b. 4 and 5

   c. −9 and 0

   d. -11 and -12

19. Write an inequality to represent: Five is no more than three less than a number.

   a. $X - 3 \leq 5$

   b. $X - 5 > 3$

   c. $X - 3 > 5$

   d. $5 \leq x - 3$

20. The quotient of a number and 3 is no greater than 15. One possible solution for the inequality is?

   a. 60

   b. 75

   c. 90

   d. 9

CHAPTER 6: AREA

6.0 Review Notes on Area of Polygons

- A polygon: closed figure that has at least three straight line segments.
- Regular polygon: polygon with all sides having equal side length
- A quadrilateral: a polygon with four sides
- A parallelogram: a quadrilateral with opposite sides parallel and opposite sides equal.
- A rectangle: a parallelogram with all of its angles 90°.
- A rhombus: a parallelogram with four equal sides.
- A square: a parallelogram with four equal angles and four equal sides.
- A trapezoid: a quadrilateral that has only one pair of parallel sides.
- Right angle: angle measure of 90°
- Acute triangle: all the three angles of the triangle measure less than 90 degree
- Obtuse triangle: one of the angles of the triangle is greater than 90°
- Right triangle: a triangle with one of the angles is 90°.
- Scalene triangle: a triangle with all angles different
- Isosceles triangle: two of the three angles or sided are the same.
- Equilateral triangle: all three sides and angles are equal
- Perimeter: distance around the figure
- Three dimensional figures: Solid figure with three-dimension, length, width, and height.
- The sum of the angles of a triangle is 180°
- The sum of the angles of a quadrilateral is 360°
- Area of a triangle = $\dfrac{base \times height}{2}$
- Area of rectangle, parallelogram = $base \times height$
- Area of square = s^2 where s is the length of the side
- Area of trapezoid = $\dfrac{(base1 + base2) \times height}{2}$

- Area of a circle, $A = \pi r^2$, r is the radius of the circle
- Circumference of a circle, $C = 2\pi r$

6.1. Properties of polygons

Say True or False

1. All quadrilaterals are parallelograms _____

2. All squares are quadrilaterals _____

3. A kite is a rhombus _____

4. A rhombus is a square _____

5. A square is a rhombus _____

6. A parallelogram is a rectangle _____

7. All trapezoids are parallelograms _____

8. A quadrilateral is a trapezoid _____

9. A square is a rectangle _____

10. Every square is a polygon _____

11. A parallelogram is a quadrilateral _____

12. Every triangle is a right triangle _____

13. An equilateral triangle is an isosceles triangle _____

14. A trapezoid is a quadrilateral _____

15. Every quadrilateral is a trapezoid _____

16. Trapezoid is a kite = _____

17. All rectangles are squares = _____

152

18. Every triangle is a rhombus = _____

19. Rhombus is a parallelogram = _____

20. All rectangles are quadrilateral = _____

21. Every square is a kite= _____

22. Equilateral triangle is quadrilateral = _____

6.2 Area of Parallelograms

Find the area of the parallelograms

1. Height = 4m and length = 5m.

2. Height=7ft.   5ft

3. Parallelogram with length 3ft. and height 2ft._____

4. Find the height of a parallelogram if the base is 20 ft. and its area is 600 ft^2.
_____

5. Find the base of a parallelogram with an area of 624 square meters and height 26 meters_____

6. Find the area of a parallelogram with base 5/4 yards and height 8 yards.
   _____

7. Find the area of a rectangle with length 6m and width 5m. _____

8. The area of a square with side length of 13in is_____.

9. Area of a rectangle is 24 ft² and the length is 6ft. its width = _____.

10. Square with area 81 yd². Length of the side = _____.

11. A rectangle has base 3m and height 4m. A parallelogram has length 3m and width 4m. Which has the smaller area? Why? _____

12. The area of a parallelogram is 5 times the area of a square with side length of 3ft. What is the area of the parallelogram? _____

13. Find the area of a parallelogram that has the same base and height as a triangle. The area of the triangle is 10 square feet. _____

14. A rectangular room measures 10m by 8m. What is the area of this room?
    _____

15. The perimeter of a rectangular playground is 46 m. If the length of the park is 7 m, what is the area of the park? _____

6.3. Area of Triangles

1. Find the area of a triangle with base 7m and height 2m. _____

2. Find the height of a triangle with area 28m² and base 8m.

3. The base and height of the triangle are doubled. Describe the change in the area compared to the original area before doubling the sizes.

4. Suppose both the base and the height are multiplied by ¼. Describe the change in the area the triangle.

5. Find the area of the right triangle with base 4m and height 3m.

6. In question 5, if each side length is doubled. Describe the change in perimeter.

7. In question 5, if each side length is doubled. Describe the change in area.

8. Find the area of an equilateral triangle with side length 4ft. and height 3.46ft

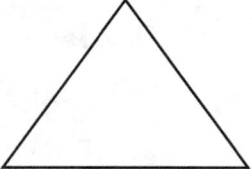

9. The area of a triangle with base 20in. is 80 in². Find the height of the triangle.

10. What are possible perimeters of an isosceles triangle with two of its side lengths 4ft and 6ft?

6.4. Area of Trapezoids

1. Find the area of a trapezoid with bases 12 cm. and 14cm. and height 8cm. _____

2. Calculate the perimeter of a trapezoid with side lengths 6, 7, 8, and 9 inches.

3. The area of a trapezoid with bases 10m and 13m is 115m². Find its height.

4. The second base of a trapezoid is twice of the first base. The area of the trapezoid is 24 square feet and the height is 8 feet. Find the two bases.

5. Find the area of a trapezoid with bases 4yd. and 6yd. and height 4yd.

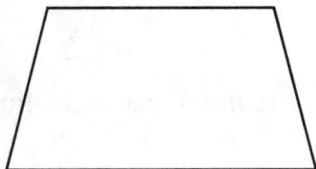

6. The two bases of a trapezoid are 4m and 3m. The height is 50cm. Find the area in square meter.

7. The area of a trapezoid is 36 ft². The two bases are 36 in. and 2ft. Find the height in inches.

8. In question 7, what is the area of the trapezoid in square inches?

9. If the two bases of a trapezoid are doubled. Describe the change in the area of the Trapezoid.

10. Describe the change in the area if the bases and the height of a trapezoid are tripled.

6.5. Area of Circles

1. What is the relation between the diameter (d) and radius (r) of the circle? Use $\pi = 3.14$

   Answer = _____

2. Find the radius of a circle if the diameter is 20 inches.

   Answer = _____

3. What is the diameter of a circle whose circumference is 31.4m?

   Answer = _____

4. Find the diameter of a circle whose radius is 5 cm?

Answer = _____

5. Find the circumference of a circle if the diameter is 10ft.

Answer = _____

6. Find the circumference if the radius is 2 units

Answer = _____

7. Circle A has a diameter of 10m. Circle B has radius of 5m. Do the two circles have the same area?

Answer = _____

8. What is the approximate value of π rounded to the nearest hundredths digit? Do we need π to calculate the area of a circle?

Answer = _____

9. Find the area of a circle that has radius of 8 yards

Answer = _____

10. What is the area of a circle whose diameter is 18 ft.?

Answer = _____

11. The area of a circle is 78.5 ft². Calculate the radius?

Answer = _____

12. The area of a circle is 25π m². What id its diameter?

Answer = _____

13. The circumference of a circle is 20π in. Find its area

Answer = _____

14. If the area of a circle is 100π square units. What is its circumference?

Answer = _____

15. What is the radius of a circle if the area is π or 3.14 square units?

Answer = _____

6.6. Area of Composite figures

1. A triangle is on the top of a rectangle sharing the same side. The length and the width of the rectangle are 5m and 6m, respectively. The triangle has a height of 4m and sharing the width with the rectangle. Find the combined area.

2. The longer side is 12m. The shorter side 6m. Find the Area.

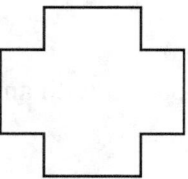

3. Find the area the combined figures. The length and the width are 10ft and 6ft.

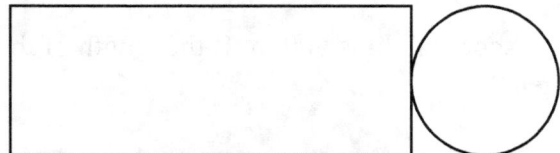

4. Find the area of the figure. Each long side = 6in. Each short side = 3in.

5. Find the area of the shaded region in the composite figure. The length and width of the rectangle are 8cm. and 5cm.

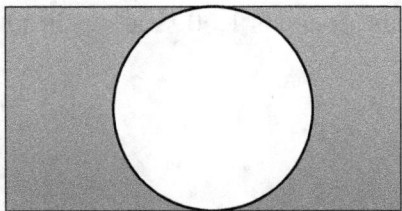

6. Find the area. The height of the triangle = 2m. The length and width of the rectangle are 6m and 4m.

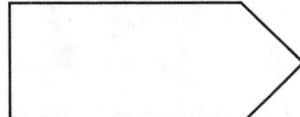

160

6.7. Perimeter and Area word problems

1. What is the height of a parallelogram with an area of 66 square feet and a base of 12 feet?

   Answer = _____

2. The perimeter of a rectangular soccer field is 500 m. If the length of the field is 100 m, what is the width of the field?

   Answer = _____

3. A rectangular bedroom is 11 m by 7.5 m. find it perimeter.

   Answer = _____

4. The perimeter of a square fence is 1,600 inches. Find the length of the side of the fence.

   Answer = _____

5. A rectangular room measures 22 m by 8 m. What is the area of this room?

   Answer = _____

6. The area of a rectangular guest house room is 360 ft². The length of the room is 12 ft. What is its width?

Answer = _____

7. Bob wants to cover two rooms with a carpet. The first room is 4m by 6m, the second room is 5m by 5m. Calculate the total area covered by the two rooms?

Answer = _____

8. In question 6, it costs Bob $12 to cover 1 square meter. How much does it cost him to cover the two rooms?

Answer = _____

9. The perimeter of a rectangular table is 28ft. The length of the table is 10 ft. find the width.

Answer = _____

10. In question 8, Jenny bought a table mat that fits the table for $180. How much is the cost of 2 ft^2 of the table mats?

Answer = _____

11. The perimeter of a rectangular track in a YMCA is 200 m. John walked twice, jogged 3 times, and then run 5 times around the track. Find the total distance covered by John in the YMCA.

Answer = _____

12. A triangle is cut from a piece of paper. The triangle has a height of 6 inches and an area of 60 in². Find the length of the base of the triangle.

   Answer = _____

13. A tile is in the form of a trapezoid. The two bases have length of 4m and 6m. The height of the trapezoid is 5m. What is the area?

6.8 Test

1. The perimeter of a rectangular table is 30ft. The width of the table is 4 ft. The height is:
    a. 7.5ft.
    b. 11ft.
    c. 6ft.
    d. 9ft.
2. A triangle is cut from a piece of paper. The triangle has a height of 6 inches and area of 60 in². Find the length of the base of the triangle.
    a. 10in.
    b. 22in.
    c. 20in.
    d. 15in.
3. A tile is in the form of a trapezoid. The two bases have length of 4m and 6m. The area of the trapezoid is 25m². What is the height?
    a. 5m.
    b. 6m.
    c. 7m.
    d. 9m.

4. Find the area of the shaded region if the bigger rectangle is 6in. by 8in. and the smaller rectangle is 5in. by 7in.

a. 11in².
   b. 10in².
   c. 12in².
   d. 13in².
5. The radius of a circle is 2.5ft. Using π = 3.14, find its area.
   a. 17.26ft²
   b. 19.26ft²
   c. 19.65ft²
   d. 19.625ft²
6. The circumference of a circle is 20π in. Find its radius.
   a. 5in.
   b. 10in.
   c. 15in.
   d. 7in.
7. If the area of a circle is 100π square units. What is its circumference?
   a. 10π units
   b. 15π units
   c. 20π units
   d. 25 units
8. Find the area of a triangle with base 3.5m and height 4m.
   a. 10m²
   b. 7m²
   c. 12m²
   d. 14m²
9. Find the height of a triangle with area 64m² and base 8m.
   a. 16m
   b. 15m
   c. 14
   d. 12m
10. Find the area of a parallelogram with base 1/4 yards and height 56 yards.
    a. 10 yards
    b. 9 yards
    c. 14 yards
    d. 7 yards
11. The height of a rectangle with area of 16m² and width 5m is:
    a. 3m.
    b. 3.1m.
    c. 3.2m.
    d. 3.3m
12. A kite is parallelogram.

a. True

b. False

c. Sometimes true

13. Every right triangle is a triangle.

    a. True

    b. False

    c. Sometimes true

14. An isosceles triangle is an equilateral triangle.

    a. True

    b. False

    c. Sometimes true

15. A trapezoid is a quadrilateral.

    a. True

    b. False

    c. Sometimes true

16. Square with an area of 81 yd² has a length of the side = _____.
    a. 7 yd.
    b. 8 yd.
    c. 9 yd.
    d. 10 yd.

17. A rectangle has base 3m and height 4m. A parallelogram has length 3m and height 4m. Which has the greater area?
    a. The rectangle
    b. The parallelogram
    c. Both have equal area
    d. We cannot compare

18. A _____ is quadrilateral with four equal sides

a. Rhombus

b. Trapezoid

c. Rectangle

d. Kite

19. Find the perimeter if the longer side is 4m and the shorter side is 2m.

a. 18m.

b. 17m.

c. 16m.

d. 32m.

20. The base and the height of a triangle are doubled. What is the change in the area of the triangle?

    a. The area will double
    b. The area will triple
    c. The area will quadruple
    d. The area will not change

CHAPTER 7: SURFACE AREA AND VOLUME

7.0. Review Notes on Surface Area and Volume

- Volume of a cube, $V = s^3$, s is the length of the side

- Volume of a rectangular prism, V = length × width × height

- Volume of a prism = B × h, B is base area and h is height of the prism.

- Volume of a sphere, $V = (4/3)\pi r^3$

- Volume of a cylinder, $V = \pi r^2 h$, where h is the height and r is the radius of the cylinder.

- Surface area = the sum of all areas of the surface of a three dimensional figure.

- Lateral surface area is a surface are excluding the bases.

- Surface area of a cube = $6s^2$, where s is the length of the side of a cube.

- Surface area of a rectangular prism = 2 × (l × w + l × h + w × h), where l is the Length, w is the width, and h is the height of the rectangular prism.

- Surface area of a sphere = $4\pi r^2$

- Surface area of a cylinder = $2\pi rh + 2\pi r^2$

7.1. Surface Area

Find the surface area of:

1. A cube with side length 3 yards

2. Rectangular prism with length, 6m, width, 4m, and height of 5m

3. Square prism with base area 25 in² and height 4 in.

4. The perimeter of a base of regular pentagonal prism is 100ft and the height is 5ft. What is the lateral surface area?

5. Find the surface area that has dimensions 2m, 4m, and 8m.

6. Find the lateral surface area of a cylinder with diameter 4ft. and height 5ft.

7. Find the surface area for question 6.

8. Nancy is painting the wall and the ceiling of her bedroom. She is not painting two windows and a door in the bedroom. Each window is 1m. by 1.5m. The door is 1m. by 2m. Her bedroom has length of 4m, width of 5m, and height of 3.5m. Calculate the surface area that Nancy is painting.

9. If Nance wants to include the windows and the door. What will be the surface area?

10. In question 8, if one gallon of paint covers 10 square meters. How many gallons of paint does Nancy need to buy?

11. Find the surface area of rectangular prism open at the top. The base has length of 5m and width of 8m. The height is 2m.

12. Find the top and bottom area of a right triangular prism. The base and height of the triangle are 4in. and 6 in.

13. Find the surface area of a cube. Each face of a cube has an area of $16m^2$.

14. Find the surface area of a cylinder with radius 6m and height 5m.

7.2. Three Dimensional Figures

Determine one possible three-dimensional figure that can be made using:

1. A square and four equilateral triangles.

2. Three rectangles and two equilateral triangles.

3. Six squares

4. Four equilateral triangles

5. Five squares and one regular pentagon.

6. Two equal circles and a rectangle.

7.3. Volume

Find the volume for each question.

1. Find the volume of rectangular prism open at the top. The base has length of 10m and width of 4m. The height is 5m.

2. Find the volume of right triangular prism. The base and height of the triangle are 4 and 6 inches, respectively. The width of the prism is 10in.

3. Find the volume of the square pyramid. The area of the base is $36m^2$ and the height is 8m.

4. A cylinder with radius 6m and height 5m.

5. A cube with side length of 8 ft.

6. Rectangular prism with length = 7in., width = 2in. and height = 4 in.

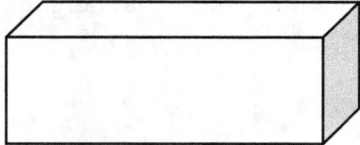

7. Base area 20cm² and height 1cm

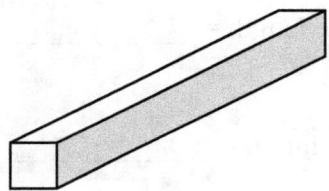

8. A square prism with base side length 4yd. and height of 3yd.

9. The radius and the height of the cylinder are 2in. and 8in., respectively. The length of the side of the cube is 5in. Find the combined volume assuming the cylinder is on the top face of the cube.

10. What is the change in volume if each side of a cube is doubled?

7.4 More on Volumes

Calculate volume or missed side.

1. Rectangular prism: Length = 4mm, width = 3mm, and height = 6mm, Volume (V) = _____

2. Cube: side length = 5in. Volume = _____

3. Rectangular Prism: Length = 4ft, width = 5ft, Volume = 60ft^3, height = _____

4. Cube: Volume = 216m^3, side length = _____

5. Surface area of a cube = 54in^2, side length = _____

6. Surface area of a cube = 96 yd^2, volume = _____

7. A rectangular pyramid: Base area = 28 m^2, and height = 6 m. Volume = _____

8. Cylinder: radius = 5in, height = 8in. Volume = _____

9. Cylinder: Volume = 36πyd^3, radius = 3. Height = _____

10. Sphere: radius = 4cm. Volume = _____

7.5. Surface area and volume word problems

1. What is the surface area of walls and the ceiling of a room that has length 20ft. width 18 ft., and height 12ft.?

   Answer = _____

2. In question 1, if it costs $6 to paint an area of 1 square foot, how much do you spend to paint the room?

   Answer = _____

3. A water tank, in the shape of a right rectangular prism, is 10 inches long, 5.5 inches wide and 4 inches high. Determine the amount of water the tank holds?

   Answer = _____

4. A rectangular prism has a base that measures 3 centimeters by 4 centimeters and unknown height. What is the height if the volume that fills the prism is 120-centimeter cubes?

   Answer = _____

5. Mark is wrapping a box that measures 7 in. by 4 in. by 5in. How much wrapping paper will Mark need to wrap the box?

   Answer = _____

6. You want to cut a paper that has an area of 2500 in^2 into small gift wrappers. How many gift wrappers can you make if each gift wrapper has an area of 25 in^2?

Answer = _____

7. A container is made in the shape of a rectangular prism. The base is 12 ft. by 3 ft. The height is 10 ft. How many books can you put in this container if you can put two books in 1 ft^3?

Answer = _____

8. If you can fill up 16 liters of water into a bathtub in one minute. How long will it take to fill this bathtub with 320 liters?

Answer = _____

9. During weekdays a family drinks 1.8L of milk a day and 2.5L during weekends. How much milk will this family drink in three weeks?

Answer = _____

10. Tom can fill his car gas tank with 20 gallons of gasoline. His brother Sam needs 3 more gallons to fill up his car gas tank. Together how many liters of gas both need to fill up their cars?

Answer = _____

7.6. Chapter 7 Test

1. Name the figure

a. Cylinder

   b. Rectangular Prism

   c. Cone

   d. Triangular pyramid

2. Name the figure.

   a. cylinder

   b. rectangular Prism

   c. Cube

   d. triangular prism

3. Find the surface area of a cube of side length 7cm.

   a. 343 cm^2

   b. 294 cm^2

   c. 584 cm^2

   d. 456 cm^2

4. A _____ is quadrilateral with four equal sides and angles

   a. Rhombus

   b. Square

   c. Rectangle

   d. Kite

5. Find the area of the figure if the longer side is 6m and the shorter side is 3m.

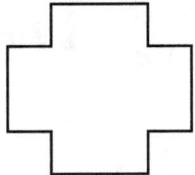

   a. 108m²

   b. 72m²

   c. 18m²

   d. 180m²

6. Find the surface area of a cube with side length of 6 yards

   a. 125 yd²

   b. 150 yd²

   c. 216 yd²

   d. 250 yd²

7. Given a rectangular prism with length = 2ft, width = 7ft, and volume = 280 ft³, the height is?

   a. 5 ft.

   b. 10 ft.

   c. 20 ft.

   d. 14 ft.

8. The volume of a cube is 27 m³, its surface area = _____

a. 10 m

b. 27 m²

c. 72 m²

d. 54 m²

9. The volume of a rectangular prism with length 10in, width 4in, and height of 20in is

   _____

   a. 60 in³

   b. 100 in³

   c. 960 in³

   d. 800 in³

10. The volume of square prism is 150 ft³ and the height is 6 ft. What is the base area of the prism?

    a. 25 ft²

    b. 50 ft²

    c. 60 ft²

    d. 30 ft²

11. Find the volume of a cylinder with radius 5cm and height 4cm?

    a. 25πcm²

    b. 10πcm

    c. 10cm

    d. 100πcm²

12. What is the surface area of a cylinder if the radius is 1in. and the height is 10 in.?

a. 12π in²

b. 22 π in²

c. 33 π in²

d. 100 in²

13. A water tank has the shape of a rectangular prism. The base is 8 ft. by 5 ft. The height is 10 ft. How many liters of water do the tank hold if 1 ft³ holds 10 liters of water?

a. 400 liters

b. 4,000 liters

c. 40, 000 liters

d. 440 liters

14. The volume of the box with length 7m, width 4m, and height 3m is

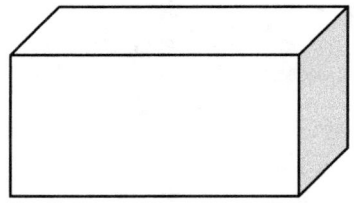

a. 72 m³

b. 24 m³

c. 84 m³

d. 96 m³

15. The area of a rectangular guest house is 4,800 ft². Each guest stays in 10ft. by 10 ft. room. How many guests can the guest house accommodate?

a. 100

b. 60

c. 80

d. 48

16. How long will it take to fill a tank with 200 liters of water if you can fill up 10 liters in half an hour?

17. Find the surface area of a square prism with base area of 25ft.^{2} and height of 10ft.

18. What is the volume of a triangular pyramid with base area of 100m^2 and height of 12m?
_____

19. The surface area of a cube is 216 in^3, find its volume _____

20. Find the length of the side of a cube that has equal volume and surface area _____

CHAPTER 8: STATISTICS

8.0 Review Notes on Statistics

- A Statistical question: a question that has more than one possible correct answers.
- Data: information collected to do statistical analysis
- Data set: a set or collection of data values

- A line plot (dot plots): using number line and dots to show data graphically. Dots are placed over the position corresponding to each data (number).

- Measures of center: describe the central value of the data set. Mean, Mode, and Median are measures of center

- Median: The middle number or the average of the two middle numbers after arranging the data in an increasing or decreasing order.

- Mean or average: sum of all values in the data divided by the number of data values.

- Mode: the most frequent value (s) in the data set

- Mean absolute deviation: the average of the difference between each data value and the mean in the data set.

- Range: the difference between the highest and lowest values in a data set.

- Distribution: shows the arrangement of data values.

- If the distribution is symmetric use the mean as a measure of center and mean absolute deviation to describe the spread. When the distribution is not symmetric use median as a measure of center and interquartile range as a measure of variation or spread.

- Gap: a number that is not a data value

- Cluster: several data values gathered in an interval

- Measures of Variation: measures how the data is scattered or spread.

- An outlier: an extremely low or high value compared to other data values.

- Five summary numbers: minimum value, first quartile, second quartile, third quartile, and the maximum value.

- Box plot: a diagram using box and line indicating the five number summary

- Frequency distribution: a table representing a date set using intervals (classes) and corresponding frequencies.

- Histogram: a bar graph representing frequency distribution

- Experiment: a process that results in possible outcomes. Ex. Flipping a coin is an experiment. The possible outcomes are head and tail.

- Outcome: result of the experiment.

- Sample space: the set of all possible outcomes of the experiment

- Event: a portion of the sample space. Sample space is a special event
- Probability of an event: number of ways the event occurs divided by the number of all possible outcomes.
- Probability can be expressed as a fraction or percent or decimal. Ex. The probability is 0.75 is the same as 75% or ¾.
- Probability is always between 0 and 1 or 0% and 100%
- Probability of 0 means impossible event.
- Probability of 1 is a certain or sure event.

8.1 Statistical Questions

Determine if each is statistical question

1. How many mails are coming in each day on your street address? _____

2. How many students think grade six math is easy? _____

3. How many people think sport is good for health? _____

4. How many states in the United States are located in the south? _____

5. How many are boys in your family? _____

6. How many branches does a tree have on it? _____

7. How many teachers are females at your school? _____

8. How many times did you have Starbucks coffee this week? _____

9. How many toys do your friends have? _____

10. How many airplanes were made yesterday in Seattle? _____

11. How many people entering a shopping mall are wearing masks? _____

12. How many books were lent in a local library yesterday? _____

13. How much money do teachers make? _____

14. How many kids does each governor in United States have? _____

15. How old is each building in Atlanta? _____

16. What did you score in yesterday's math test? _____

17. What was the students score for English test? _____

18. How old is your oldest son? _____

19. How many times did you fly last week? _____

20. How many cars does each person in your neighborhood own? _____

8.2. Measures of Center (Mean, Median, and Mode)

8.2.1 Mean

1. Find the mean score of the scores indicated in the dot plot

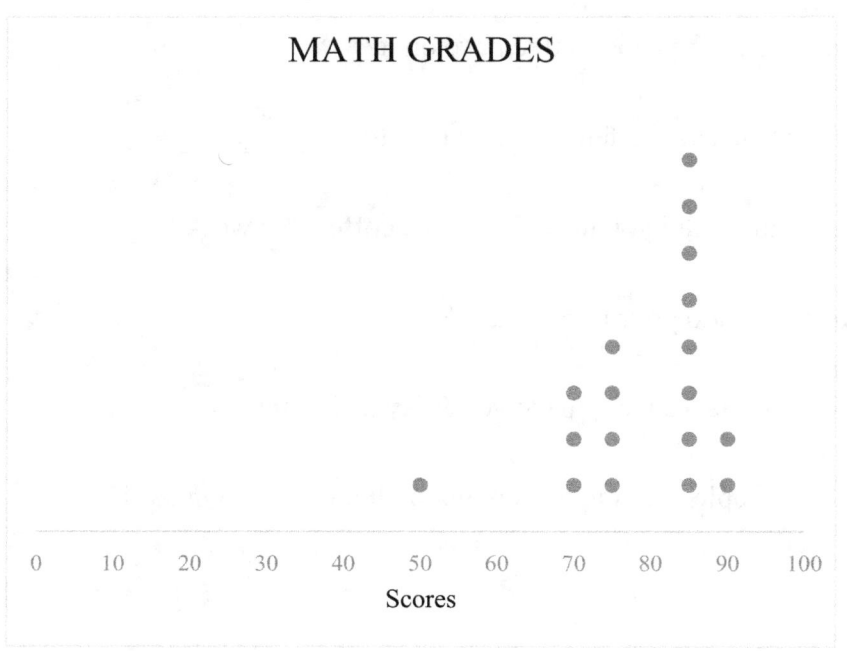

2. What is the mean of the following numbers?

a. 10, 39, 71, 39, 76, 38, 25

b. 10, 20, 30, 40, 50, 60, 60, 60, 90, 80

c. 5, 4, 10, 3, 3, 4, 7, 4, 6, 5, 11, 9, 5, 7

d. 55, 55, 55, 55, 55, 55, 55.

e. 23. 34, 45, 66, 42

f. A student recorded her scores on weekly math quizzes that were marked out of a 10 points. Her scores were: 7, 9, 8, 5, 7, 6, 7, 7, 5, 7, 5, 5, 6, 6, 9, 8, 9, 7, 9, 9, 6, 8, 6, 6, 7. Find her mean quiz score.

Answer = _____

3. What number would you divide by to calculate the mean of?

    a. 2, 3, 4

Answer = _____

    b. 20, 40, 50, 6, 4, 7, 8

Answer =_____

4. The mean of four numbers is 21.5. If three of the numbers are 18, 26, and 25, what is the value of the fourth number?

Answer =_____

5. The mean of five numbers is 10. If four of the numbers are 8, 10, 12, and 4, what is the value of the fifth number?

Answer =_____

6. The mean of six numbers is 5. If five of the numbers are 7, 3, 5, 6 and 4, what is the value of the sixth number?

Answer =_____

7. Calculate the mean weekly income of the person indicated below

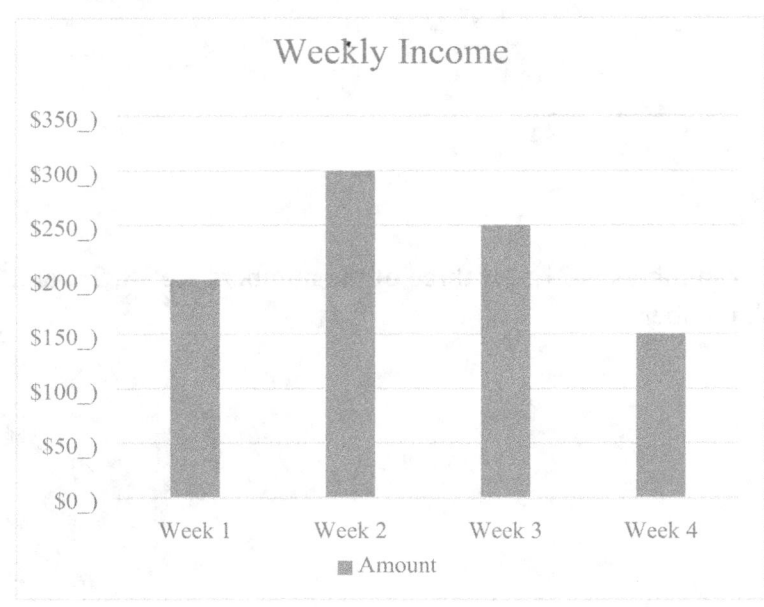

8. Mary bought 10 masks. She spent $5.2, $4.8, $4, $9, $8, $3.25, $4.5, $6.5, $5.75 for the nine masks. The average amount of money she spent was $5.5. What is the cost of the tenth mask?

9. Check if group A and group B have the same mean. Group A: 12, 14, 13. Group B: 33, 11, 8, 4

10. In question 9, if you replace the number 12 in group A by another number both groups will have the mean. What number would you replace?

8.2.2 Median

1. What is the median of the following numbers?

    a. 87, 44, 10, 39, 71, 39, 76, 38, 25

Answer =_____

b. 50, 20, 30, 40, 50, 60, 60, 60 90, 85, 45, 33

Answer =_____

c. 4, 4, 4, 4, 4

Answer =_____

d. 84, 55, 66, 42, 75, 33

Answer =_____

2. Suppose you have 13 numbers arranged in order from the smallest to the largest. In which position is the median located?

Answer =_____

3. If you have 10 numbers arranged in order from the smallest to the largest. How do you calculate the median of the numbers?

Answer =_____

4. The median of three numbers is 23. Two of the numbers are 10 and 28. What is the third number?

Answer =_____

5. The Median of four numbers is 4. Three of the numbers are 2, 3, and 7. What is the fourth number?

Answer =_____

6. Find the median of the numbers: 14, 21, 12, 12, 13, 16, 27, 35, 11, and 46 is _____.

Answer =_____

8. 2.3. Mode

Determine if the data has no mode, unimodal, or bimodal.

1. 3, 4, 6, 3, 8

    Answer = _____

2. 2, 3, 5, 4, 2, 18, 20, 20

    Answer = _____

3. 1, 2, 3, 4, 5, 6, 7, 8

    Answer = _____

4. 2, 3, 2, 3, 2, 3, 4, 5, 4.

    Answer = _____

Find the mode of the numbers

5. 23, 24, 26, 23, 2 8

   Answer = _____

6. 12, 13, 15, 14, 13, 2, 18, 20, 20

   Answer = _____

7. 51, 52, 53, 54, 55, 56, 57, 58, 52, 33, 52

   Answer = _____

8. 2, 3, 2, 3, 2, 3, 4, 5, 4.

   Answer = _____

8.2.4 Mixed Problem

2. Give an example of data having (May have more than one answer)

a. Equal mean and median

   Answer = _____

b. Equal mean and mode

   Answer = _____

c. Equal mode and mean

   Answer = _____

d. Equal mean, mode and median

Answer = _____

8.3. Measures of Variation. (Range, Mean absolute deviation, Quartiles and Interquartile Range)

8.3.1. Range

Find the range of the numbers

1. 6, 83, 1, 2, 4, 6, 9. 0.5

    Answer = _____

2. 10, 10, 10, 10, 10

    Answer = _____

    3. 42, 42, 65, 67, 65, 87, 19

    Answer = _____

    4. 25, 125, 32, 428, 32, 511, 34, 287, 36, 47, 89, 40, 302, 38, 256

    Answer = _____

    5. The range of the numbers: 7, 8, ?, 5, and 6 is 10. The missed number "?" is _____

6. The range of the numbers: 3, 8, ?, 5, and 6 is 6. One possible value for the missed number "?" is _____

7. In question 6, what is the second possible value for the missed number "?" is _____

8. Give an example of list of three numbers that have a range of 5 (answer varies).

   Answer = _____

9. Give an example of list of four numbers that have a range of 0 (answer varies).

   Answer = _____

10. Give an example of list of four numbers that have a range of 1 (answer varies).

   Answer = _____

8.3.2 First, Second, and Third Quartiles

Find the 1st, 2nd, and 3rd quartile of the numbers

1. 7, 1, 4, 6, 5, 2. 3 ____, ____, _____

2. 12, 3, 4, 5, 9 ____, ____, _____

3. 5, 8, 6, 2, 3, 9 ____, ____, _____

4. 3, 9, 1, 3, 5, 2 ____, ____, _____

5. 10, 2, 8, 6, 9, 5, 5____, ____, _____

6. 86, 85, 22, 46, 61, 32, 39, 22, 75, 33, 86____, ____, _____

7. 2, 7, 6, 3, 1, 5, 7, 8____, ____, _____

8. 9, 4, 1, 9, 9, 3, 7, 5____, ____, _____

9. 1, 6, 1, 5, 8, 2, 3, 7____, ____, _____

10. 78, 74, 45, 35, 68, 45, 45, 63, 73, 85, 49____, ____, _____

11. 72, 95, 38, 37, 45, 54, 71, 23____, ____, _____

12. 22, 48, 88, 75, 60, 30, 53, 92, 67, 77____, ____, _____

13. 83, 23, 87, 81, 28, 69, 42, 24____, ____, _____

14. 24, 81, 26, 22, 28, 25, 68, 30____, ____, _____

15. 13, 45, 21, 12 ____, ____, _____

16. 4, 4, 4, 4 ____, ____, _____

8.3.3 Interquartile Range

Find the interquartile range of the numbers.

1. 7, 1, 4, 6, 5, 2. 3 _____

2. 12, 3, 4, 5, 9_____

3. 5, 8, 6, 2, 3, 9_____

4. 3, 9, 1, 3, 5, 2_____

5. 10, 2, 8, 6, 9, 5, 5_____

6. 86, 85, 22, 46, 61, 32, 39, 22, 75, 33, 86 _____

7. 2, 7, 6, 3, 1, 5, 7, 8 _____

8. 9, 4, 1, 9, 9, 3, 7, 5 _____

9. 1, 6, 1, 5, 8, 2, 3, 7 _____

10. 78, 74, 45, 35, 68, 45, 45, 63, 73, 85, 49 _____

11. 72, 95, 38, 37, 45, 54, 71, 23 _____

12. 22, 48, 88, 75, 60, 30, 53, 92, 67, 77 _____

13. 83, 23, 87, 81, 28, 69, 42, 24 _____

14. 24, 81, 2622, 28, 25, 68, 30 _____

15. 13, 45, 21, 12 _____

16. 4, 4, 4, 4 _____

8.3.4. Mean absolute deviation.

Calculate the Mean absolute deviation of the numbers rounded to one decimal place.

1. 7, 1, 4, 6, 5, 2, 3 _____

2. 12, 3, 4, 5, 9 _____

3. 5, 8, 6, 2, 3, 9 _____

4. 7, 1, 19, 1_____

5. 4, 5, 16, 7_____

6. 3, 16, 12, 21, 13_____

7. 25, 3, 22, 14, 16_____

8. 5, 11, 12, 12, 7, 13_____

9. 6, 12, 13, 13, 8, 14_____

10. 78, 74, 45, 35, 68, 45, 45, 63, 73, 85, 49_____

8.4 Outlier

Find the minimum, the maximum value for the numbers. Determine if there is an outlier.

1. 7, 1, 19, 1_____, _____, _____

2. 4, 5, 16, 7_____, _____, _____

3. 3, 16, 12, 21, 13_____, _____, _____

4. 25, 3, 22, 14, 16_____, _____, _____

5. 5, 11, 12, 12, 7, 13_____, _____, _____

6. 6, 12, 13, 13, 8, 14_____, _____, _____

7. 78, 74, 45, 35, 68, 45, 45, 63, 73, 85, 49_____, _____, _____

8. 72, 95, 38, 37, 45, 54, 71, 23_____, _____, _____

9. 22, 48, 88, 75, 60, 30, 53, 92, 67, 77_____, _____, _____

10. 83, 23, 87, 81, 28, 69, 42, 24_____, _____, _____

11. 24, 81, 26, 22, 28, 25, 68, 30 _____, _____, _____

12. 13, 45, 20, 21, 180 _____, _____, _____

13. 6, 6, 6, 6, 6 _____, _____, _____

8.5. Box & Whisker Plot

The box and whisker plot shows the test result of a driving test. Answer each question based on the graph

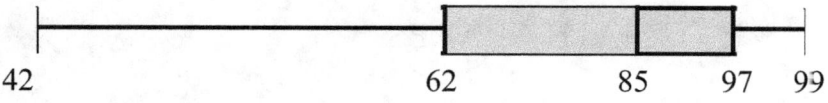

1. What is the lowest test score? _____

2. What is the highest test score? _____

3. What is the 1st quartile? _____

4. What is the 3rd quartile? _____

5. What is the interquartile range? _____

6. Is 42 an outlier? _____

7. What percent of the class scored below 62? _____

8. What percent of the class scored between 42 and 85? _____

9. What percent of the class scored above 62? _____

10. What percent of students scored below 97? _____

8.6. Line plots

Answer the questions based on the line plot below

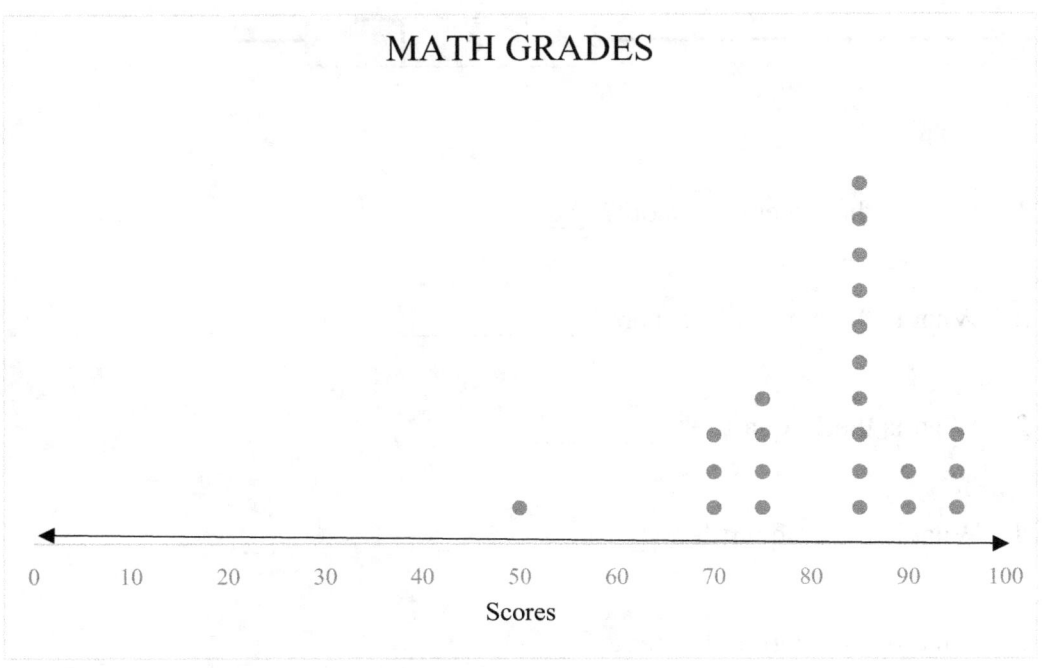

1. How many students got 70? _____

2. What was the highest score in the class? _____

3. What was the lowest score in the class? _____

4. How many students received a score in the 70s? ____

5. How many students received a score in the 80s? ____

6. How many students received a score in the 60s? ____

7. How many students scored 85% or less? _____

8. How many students scored 85% or above? _____

9. How many students are in the class? _____

10. How many students scored between 63 and 84? ____

11. What score is received by most of the students? _____

12. What percentage of students scored 85? _____

13. What percentage of students scored 90 or less? __

14. What percentage of students scored above 99? _____

15. What percentage of the students scored 50 or more____

8.7. Bar Graphs

8.7.1. Interpreting a bar graph

The bar graph below is the number of tourists visited five countries in June of 2018.

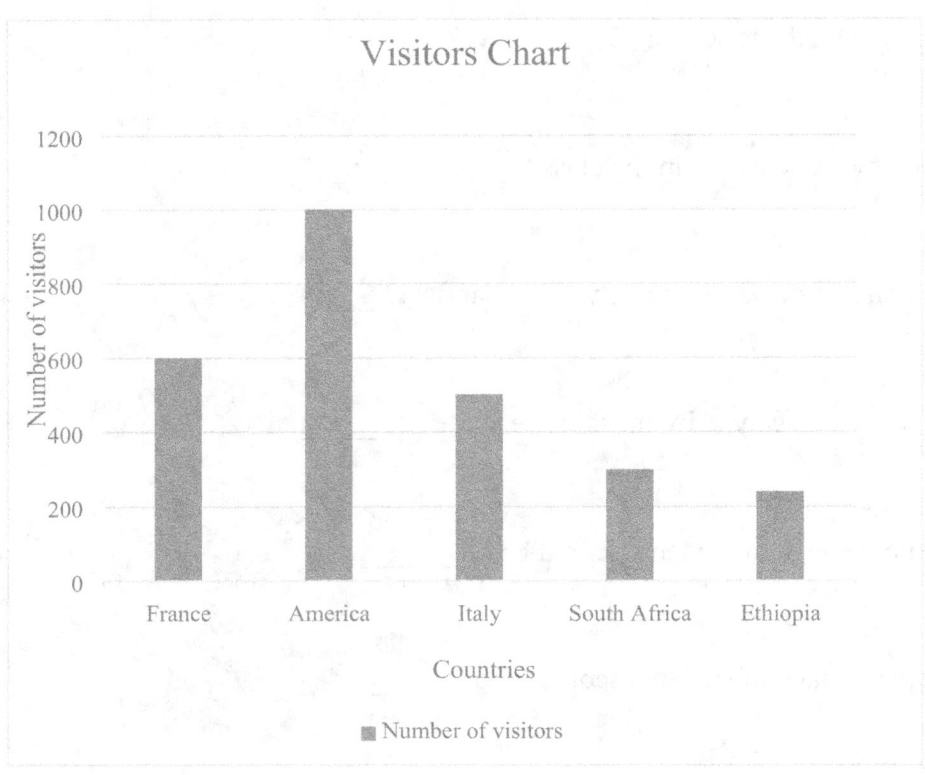

1. How many tourists visited Italy? = _____

2. Which country has the highest tourist? = _____

3. Were there more tourists in Italy than France? = _____

4. Which country has exactly 600 tourists? = _____

5. Which country has the lowest number of tourists? = _____

6. What is the combined number of visitors from Italy and America? = _____

7. What is the difference between the number of visitors in America and Italy? = _____

8. What is the total number of visitors? = _____

9. How many more tourists visited France than Ethiopia? = _____

10. How many tourists traveled to Africa? = _____

8.7.2 Constructing a bar graph

Construct a bar graph and answer the questions below.

Grade	Number of students
A	50
B	150
C	200
D	40
F	10

1. How many students received a grade of C? = _____

2. How many students received a grade of B or higher? = _____

3. How many students received a grade of B or below? = _____

4. What is the most frequent grade received? = _____

5. What percent of students got A? = _____

6. What is the total number of students? = _____

7. How many fewer students got B than C? = _____

8. How many students received a passing grade? Assume the passing grade is a grade of C or above? = _____

9. What percentage of students passed? = _____

10. Does 50% of the grade fall above C? = _____

8.8. Line graph

Answer the question based on the line graph below

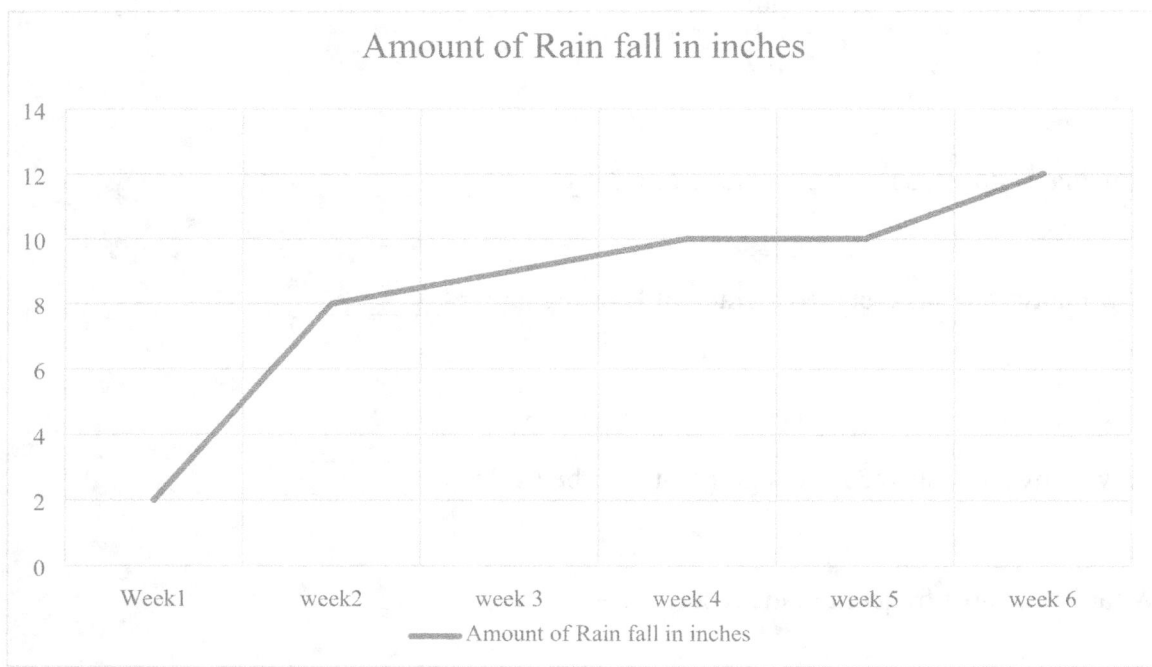

1. Which week has the lowest rainfall? _____

2. Rainfall _____ as the number of week increase (increase or decrease)

3. Which week has the highest rainfall? _____

4. What is the average amount of rainfall in week 4? _____

5. What is the range of rainfall received between week 1 and week 6? _____

6. What is the mean weekly amount of rainfall? _____

7. Which week has 12 inches of rainfall? _____

8. Which week has 8 inches of rainfall? _____

9. Which week has 9 inches of rainfall? _____

10. From the trend of the line graph can we conclude week 7 will have at least 12 inches of rainfall? _____ (yes, no)

8.9. Frequency Distribution and Histogram

8.9.1 Use the frequency distribution and answer the questions 1 – 10 below.

Temperature in Summer	Number of days
0°F – 19°F	2
20°F - 39°F	4
40°F - 59°F	6
60°F - 79°F	30
80°F - 99°F	50
100°F - 119°F	8

1. What percent of the temperature is between 80°F and 99°F degree? ___

2. How many days fall in between 20°F and 99°F? _____

3. How many days have a temperature of 20 or more? _____

4. How many days have a temperature of 59°F or less? _____

5. What is the most frequent temperature range? _____

6. What is the least frequent temperature range? _____

7. What is the possible maximum temperature? _____

8. If you pick a day in summer, in which range of temperature is the day most likely be

   _____

9. What percent of the temperature is below 79°F degree? ___

10. What percent of the temperature is above 119°F degree? ___

11. Construct frequency distribution using the following date. Use classes 0 – 9, 10 – 19, 20 – 29, 30 – 39, and 40 – 49.

2, 4, 9, 11, 5, 6, 33, 19, 23, 45, 37, 24, 33, 29, 29, 12, 26, 18, 19, 3, 5, 4, 6, 16, 26, 36, 46, 49, 22, 28, 34, 43, 11, 9, 3, 6, 11, 22, 33, 44

8.9.2 Histogram

Interpreting Histograms.

The histogram below shows the weight of adult tenants in pounds in south wood vista apartment complex. Answer each question based on the histogram.

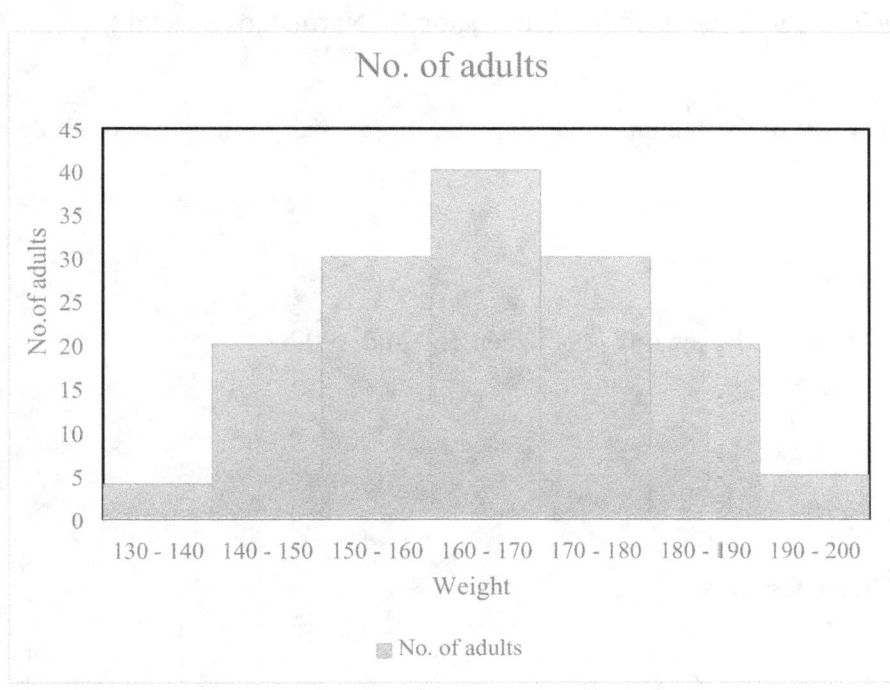

1. How many adult tenants live in the apartment? _____

2. How many adult tenants weigh 170 lb. or more? _____

3. How many adult tenants weigh 160 lb. or less? _____

206

4. How many adult tenants weigh between 140 and 200 lb.? _____

5. What is the maximum weight range? _____

6. What was the most common weight range of adult tenants at the Southwood vista apartment? _____

7. How many adult tenants were overweight? Consider 170 lb. or above as an overweight _____

8. In which category most adult tenants belong? (Normal, overweight) _____

9. What percent of the tenants is at least 160 lb. (answer to the nearest whole number)? _____

10. What percent of the tenants is between 150 and 190lb.? _____

8.10. Circle Graphs

Interpreting Circle Graphs.

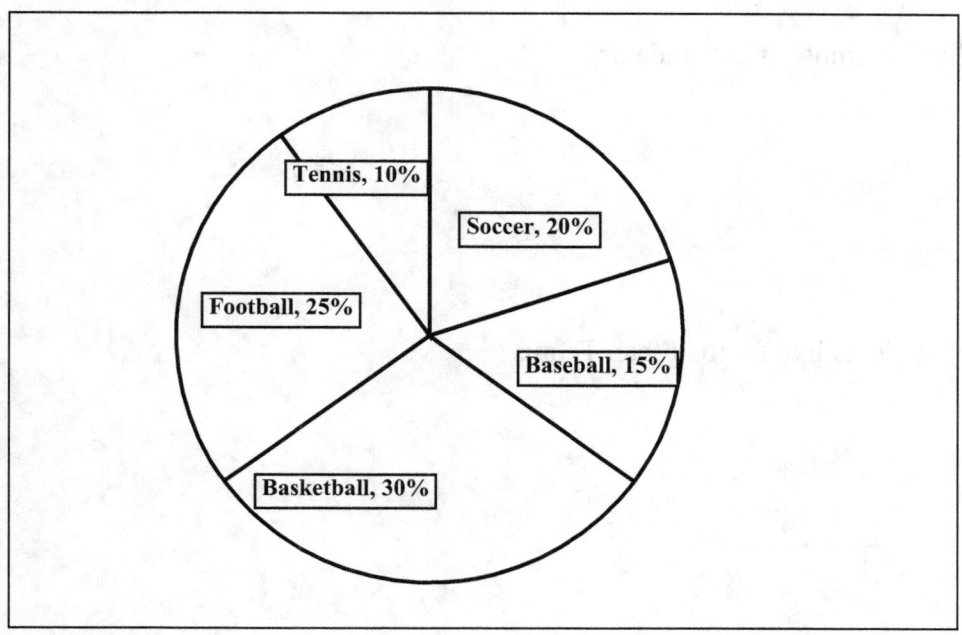

Answer the following questions based on the above Circle graph (pie chart) of 400 students' favorite sports.

1. How many students like soccer?

   Answer = _____

2. How many students like baseball?

   Answer = _____

3. How many Students like tennis?

   Answer = _____

4. Which sport is liked by most of the students?

   Answer = _____

5. How many more students like football than Tennis?

   Answer = _____

6. How many students like soccer and tennis?

   Answer = _____

7. How many students like football, soccer, and tennis?

   Answer = _____

8. How many students like sports other than basketball?

   Answer = _____

9. If you pick a student by chance, he is more likely to be basketball lover?

Answer true or false

Answer = _____

10. If you pick a student by chance, the student is less likely to be tennis lover?

Answer true or false

Answer = _____

8.11. Compared Data Set

The bar graph below is the grade distribution of male and female students at a certain School. Answer the questions based on the chart.

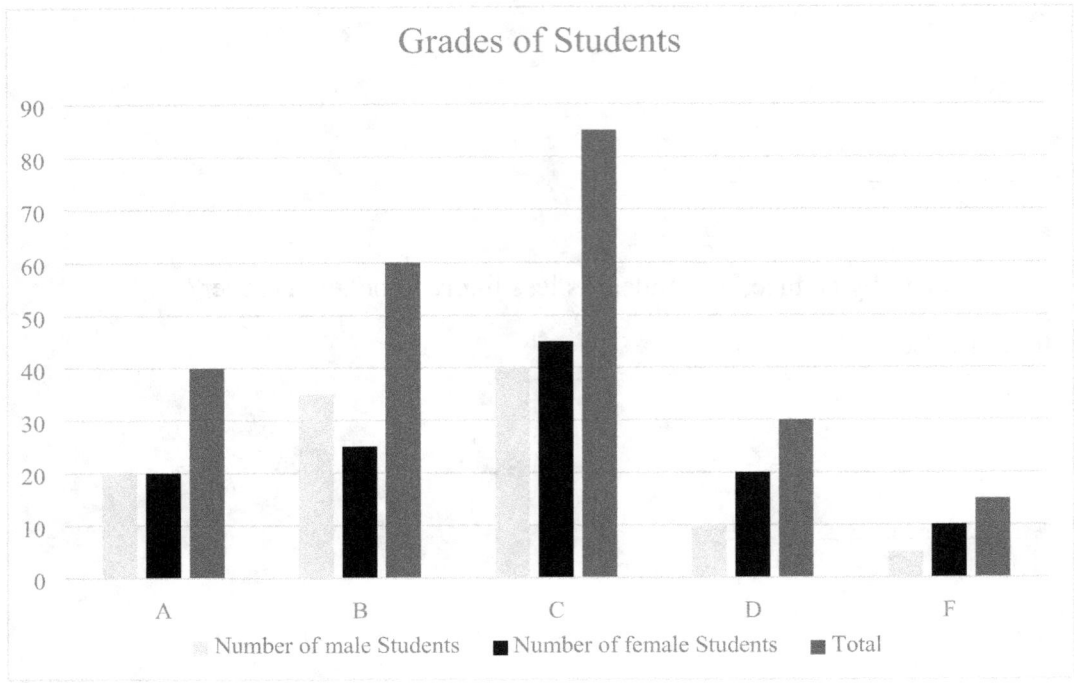

1. How many male students received a grade of B? = _____

2. How many female students received a grade of C or higher? = _____

3. How many female students received B or lower? = _____

4. What percentage of students received B? = _____

5. What percentage of female students received B? = _____

6. What percent of students got A or B? = _____

7. What is the total number of students in the school? = _____

8. How many fewer students got B than C? = _____

9. How many male students received a passing grade? Assume the passing grade is a grade of C or above? = _____

10. How many female students received a passing grade? Assume the passing grade is a grade of C or above? = _____

11. What percentage of male students passed? = _____

12. What percentage of students passed? = _____

13. What percentage of students are female? = _____

14. Does 50% of the grade fall above C? = _____

15. Comparing the number male and female students with grade of C. Males got more C than female (true or false) = _____

8.12. Basic probability properties

1. Identify the number that can be considered as probability.

a. 2, 2%, 35, 4 _____

b. -0.5, 2.4, 1.9, 12, 4/5 _____

c. 1.1, -2, -0.01, 40% _____

d. 199%, 0.89, 78, 3.4 _____

e. 66%, -0.91, 12, 121%, 145% _____

8.13. Probability Experiment & Outcomes

8.13.1 Write the probabilistic experiment.

1. Tossing a Coin

Answer = _____

2. Tossing two coins

Answer = _____

3. Rolling a die

Answer = _____

4. A spinner is divided into eight equal sections numbered one to eight. Spinning the pointer will indicate in one of the eight numbers

Answer = _____

5. Pulling a ball from a bag containing two red, one blue, three green, and two yellow balls.

Answer = _____

6. Assigning numbers 2, 3, 4, 5, 6, and 7 to six faces of a cube.

Answer = _____

7. Coloring faces of a box with red, yellow, blue, green, white, and black colors

Answer = _____

8.13.2 Write all possible outcomes

1. Turning the light switch

Answer = _____

2. Tossing two coins

Answer = _____

3. Rolling a die

Answer = _____

4. A spinner is divided into eight equal sections numbered one to eight.

Answer = _____

5. Pulling a ball from a bag containing two red, one blue, three green, and two yellow balls.

Answer = _____

6. Assigning numbers 2, 3, 4, 5, 6, and 7 to six faces of a cube.

Answer = _____

7. Coloring faces of a box with red, yellow, blue, green, white, and black colors

Answer = _____

8.14. Probability as a Percent

Write the answer in percent form. A bag contains 5 green, 3 blue, 2 red, and 10 yellow marbles. Find the probability of picking:

a. green marble = _____

b. blue marble = _____

c. red or yellow marble = _____

d. yellow marble = _____

e. green or red marble = _____

f. green or yellow = _____

g. green or red or yellow = _____

h. not green marble = _____

i. not red marble = _____

8.15. Probability as a Fraction

1. Roll a die. Write fraction for the probability of:

a. a 2 showing up = _____.

b. a 3 showing up = _____

c. a 2 or a 3 showing up = _____

d. An even number showing up = _____

e. An odd number showing up = _____

f. a 2 or a 5 showing number = _____

2. A spinner has nine equal sections colored with 3 blues, 2 reds, 2 yellows, 1 green, and 1 black. Write fraction for the probability of the spinner pointing

a. red = _____

b. yellow = _____

c. green = _____

d. black = _____

e. blue = _____

f. red or yellow = _____

g. blue or green = _____

h. blue or red or yellow = _____

i. not yellow = _____

8.16. Word problems on Statistics and Probability

1. The number of texts sent per day by 10 students are: 30, 12, 28, 200, 4, 10, 6, 2, 8 and, 20

    a. What is the 1st quartile? _____

    b. What is the 3rd quartile? _____

    c. The interquartile range? _____

    d. Do you think the student who sent 200 texts is an outlier? _____

    e. What is the median number of texts sent? _____

    f. What is the range? _____

    g. What is the best measure of center for the data? _____

    h. What is the best measure of spread for the data? _____

    i. What is the mean absolute deviation? _____

2. The ages of 25 soccer players are: 18, 19, 19, 20, 20, 20, 21, 21, 21, 21, 22, 22, 22, 22, 22, 23, 23, 23, 23, 24, 24, 24, 25, 25, 26. Make a dot plot and answer each question.

    a. What is the 1st quartile? _____

    b. What is the 3rd quartile? _____

    c. The interquartile range? _____

    d. Do you think 26 is an outlier? _____

    e. What is the median age? _____

    f. What is the range? _____

    g. What is the best measure of center for the data? _____

    h. What is the best measure of spread for the data? _____

i. What is the pick of these numbers? _____

j. Do you see gap in the data set? _____

3. The height of 10 students measured in feet are: 6.75, 5.75, 6.25, 4, 5.5, 5.5, 6.5, 5.25, 5.5, 7. Answer the following question based on the given data.

   a. What is the difference between the tallest and shortest student? _____

   b. What is the most common height? _____

   c. How many students are shorter than 6.5 feet? _____

   d. What is the average height of the students? _____

   e. What is the median height? _____

   f. What percent of students are taller than 5.75 feet? _____

   g. How many students have heights between 4.1 feet and 6.9 feet? _____

   h. Percentage of students' taller than 7.2 feet? _____

   i. Percentage of students' greater than or equal to 4 feet? _____

   j. Is the range of the heights greater than 3.5 feet? _____

Find the probability of each.

4. Turn the light switch. The chance that the light is on = _____.

218

5. Toss a coin. The probability that heads showing up = _____.

6. Rolling a die. The probability that an odd number shows up = _____.

7. A spinner is divided into eight equal sections numbered one to eight. Spinning the pointer will indicate in one of the eight numbers. The probability that the pointer indicates a number greater than two = _____.

8. Pull a ball from a bag containing two red, one blue, three green, and two yellow balls. The probability that the ball is not green = _____.

9. Assigning numbers 1, 2, 3, 4, 5, and 6 to six faces of a cube. The chance that the number four shows up = _____.

10. The faces of a cube are colored with red, yellow, blue, green, white, and black colors. The probability that a blue or green colored face shows up = _____.

11. Sally took a math test containing 40 questions. How many questions should she answer correctly to make 80% = _____?

12. If you buy 8 of 100 tickets to win a prize, what is the chance of winning the prize = _____?

13. A coin is tossed and a ball is picked from a bag containing three balls colored blue, red and black. If you have to pick a coin and a ball, what is the chance to pick heads faced coin and blue ball = _____?

8.17. Chapter 8 Test

The number of books owned by seven family members are: 1, 2, 5, 9, 11, 12, and 50. Answer question 1 – 5.

1. The 1st quartile is _____.

    a. 2
    b. 3
    c. 4
    d. 5

2. The interquartile range is _____

    a. 8
    b. 9
    c. 10
    d. 11

3. Is 50 an outlier?

    a. Yes
    b. No

4. The mean absolute deviation is _____

    a. 4
    b. 8
    c. 7.7
    d. 10.6

5. What is the best measure of variation for the data?

    a. Range
    b. Interquartile range
    c. 3rd quartile
    d. Mean absolute deviation

6. What is the range of the following numbers? 20, 39, 71, 39, 86, 38, 25

    a. 42

b. 39

c. 66

d. 35.5

7. What number would you divide by to calculate the mean of 1, 3, 4, 5, and 6?

   a. 6

   b. 3

   c. 5

   d. 4

8. What measure of central tendency is calculated by adding all the values and dividing the sum by the number of values?

   a. Median

   b. Mean

   c. Mode

   d. Typical value

9. The mean of four numbers is 71.5. If three of the numbers are 58, 76, and 88, what is the value of the fourth number?

   a. 64

   b. 60

   c. 76

   d. 82

10. The mean of the following set of numbers: 50, 51, 85, 69, 19, 60, 80, 63, 109, 42

    a. 55

b. 53.3

c. 43

d. None

11. The mean weight of five adult males is 167.2 pounds. The weights of four of the males are 158.4 pounds, 162.8 pounds, 165 pounds, and 178.2 pounds. What is the weight of the fifth male?

   a. 200

   b. 171.6

   c. 186.4

   d. None of the above

12. The mean width of 12 books is 8.1 inches.
What is the total width of the books? _____

13. The following data represent the number of pop-up advertisements received by 10 families during the past months. Calculate the range of pop-up advertisements received in the past months _____

43   37   35   30   41   23   33   31   16   21

14. A group of customer service surveys were sent out at random.
The scores were 90, 50, 70, 80, 70, 60, 80, 30, 80, 90, and 20. Find the modal score.

_____

15. What is the median of the following numbers?
10, 39, 71, 42, 39, 76, 38, 25

   a. 42.5

   b. 39

   c. 42

   d. 35.5

16. The front row in a movie theatre has 23 seats. If you were asked to sit in the median position, in which seat would you be seating?

    a. 1

    b. 11

    c. 23

    d. 12

17. What is the median score achieved by a student who recorded the following scores on 10 math quizzes? 68, 55, 70, 62, 71, 58, 81, 82, 63, 79

    a. 68

    b. 71

    c. 79

    d. 69

18. A set of four numbers that begins with the number 32 is arranged from smallest to largest. If the median is 38, which of the following could possibly be the set of numbers?

    a. 32, 32, 36, 38

    b. 32, 35, 38, 41

    c. 32, 34, 36, 35

    d. 32, 36, 40, 44

19. The number of service upgrades sold by each of the 9 employees are:

32, 6, 21, 10, 8, 11, 12, 36, 17

What is the median number of service upgrades sold by the 9 employees? _____

20. Which of the following measures can be determined for quantitative data?

    a. Mean

    b. Median

    c. Mode

d. All of these

21. What is the term used to describe the distribution of a data set with one mode?

    a. Multimodal

    b. Unimodal

    c. No mode

    d. Bimodal

22. What is the mode of the following numbers? 12, 11, 14, 10, 8, 13, 11, 9

    a. 11

    b. 10

    c. 14

    d. 8

23. Which of the following measures can have more than one value for a set of data?

    a. Median

    b. Mode

    c. Mean

24. What is the range of the following sets of numbers? 12, 0, 15, 15, 13, 19, 16, 13, 16, 16

25. A student recorded her scores on weekly math quizzes that were marked out of 10 points. Her scores were: 8, 5, 8, 5, 7, 6, 7, 7, 5, 7, 5, 5, 6, 6, 9, 8, 9, 7, 9. How many quizzes did she take? _____

26. Which of the numbers can be the probability of a certain event.

    a. 3

    b. 0.1

    c. 9.3

    d. -0.22

27. Roll a die and see the number facing up. The probabilistic experiment is

    a. Tossing a coin

b. spinning a spinner

c. Rolling a die

d. pulling a card

28. Toss a coin. The possible outcomes of this experiment is:

   a. 1, 2, 3, 4, 5, and 6

   b. Heads, tails

   c. Heads

   d. Red, Yellow

29. A couple have a child. The possible genders are:

   a. Boy, Girl

   b. Boy, Boy

   c. Girl, Girl

   d. Boy, Boy, Girl

30. What is the probability of a number 5 showing up in rolling a die?

   a. ¾

   b. ½

   c. 5/6

   d. 1/6

31. What is the probability of a 3 or a 4 showing up in rolling a die?

   a. 1/6

   b. 1/3

   c. 2/3

   d. 5/6

32. What is the probability of an even number showing up in rolling a die?

    a. 1/2

    b. 2/3

    c. 1/5

    d. 5/6

33. A spinner has nine equal sections colored with 3 blues, 2 reds, 2 yellows, 1 green, and 1 black. Find the fraction for the probability of the spinner pointing blue

    a. 1/3

    b. 2/7

    c. 4/9

    d. 1/9

34. In question 8, what is the probability of the spinner pointing red or yellow?

    a. 2/9

    b. 3/9

    c. 4/9

    d. 5/9

35. In question 8, what percentage of the spinner points blue or red (answer round to the nearest whole number)?

    a. 30%

    b. 40%

    c. 50%

    d. 56%

Pull a ball from a bag containing two red, one blue, three green, and two yellow balls. Answer question 36 – 40.

36. The probability that the ball is red _____

37. The probability that the ball is red or blue _____

38. The probability that the ball is not blue _____

39. The probability that the ball is black _____

40. The probability that the ball is green or red or blue or yellow ___

CHAPTER 9: INTEGERS

9.0 Review Notes on Integers

- Whole numbers: numbers: 0, 1, 2, 3, 4, 5, 6 …

- Negative integers: -1, -2, -3, -4, -5, - 6 …

- Integers: whole numbers combined with negative integers.

- Integers: …, -7, -6,-5,-4,-3 , -2,-1 , 0, 1, 2, 3, 4, 5, 6, 7, …

- The order of negative numbers is opposite to the order of positive integers. For example 3 < 5 but – 3 > -5.

- The sum of two positive numbers is always a positive number.

- The sum of a positive and a negative number could be positive or negative. If the negative number is bigger, then the sum will have a negative sign. If the positive number is bigger, then the sum will be positive.

- The product of two negative integers is always a positive integer. The product of a positive and negative integer is always a negative integer. The rule for division is the same as the rule for multiplication.

- Numbers like 3 and -3, 5 and -5 are opposite to each other. The opposite of -10 is 10.

- A coordinate plane: is a two-dimensional plane formed by vertical and horizontal axes.

- X – axis: the horizontal axis in the coordinate plane

- Y – axis: the vertical axis in the coordinate plane

- A coordinate plane has four quadrants. In the first quadrant both x and y coordinates are positive. In the second coordinate x is negative and y is positive. In the third quadrant both x and y coordinates are negative. In the fourth quadrant x is positive and y is negative.

- Origin: the intersection point of the x and y axes.

- Origin has x = 0 and y = 0 numbers related to it, so (0, 0) is a pair (x and y coordinates) associated to the origin.

- A point on a coordinate plane has pair of numbers (x, y) associated with it. The first number is the x coordinate and the second number is the y coordinate.

- To locate coordinates of a point (x, y). Start from origin and move x units to the right if x is a positive or x units to the left if x is a negative. Then move up y units if y is a positive or go down y units if y is a negative.

9.1. Integers

Identify the number that is not an integer.

1. 5, 6, 0.1, 22, 45 = _____

2. 2, 22, .222, 22.2, 2000 = _____

3. ½ , 44, 3^3, 90, 13 = _____

4. 10, - 1000, 43, 50/10, 2.5 = _____

5. 79, -97, 89, -98, 9.8 = _____

6. ¾, 80, 7, 55, 66 = _____

7. 1.25, - 6, 0, 67 = _____

8. 88, π, 9, - 44, 90 = _____

9. -78, 35, 24, 1/3 = _____

10. 686, 200, - 77, 3/8, -9, - 24 = _____

9.2 Opposite Integers

Write the opposite of each integer.

1. 2 = _____

2. 5 = _____

3. -3 = _____

4. 16 = _____

5. -12 = _____

6. -33 = _____

7. -100 = _____

8. 45 = _____

9. -602 = _____

10. 9,100 = _____

9.3. Comparing Integers

9.3.1. Compare the following integers using < (less than) or > greater than or = signs.

1. -5 _____ -6
2. -10 _____ -9
3. 0 _____ -4
4. 45 _____ -43

5. 2 _____ 2

6. 20 _____ -20

7. − 4 _____ -5

8. 7 _____ -4

9. 41 _____ 40

10. -41 _____ -40

11. 3 _____ 4

12. 4 _____ -5

13. 9 _____ -9

14. -10 _____ -11

15. -15 _____ 0

16. – 11 _____ -12

17. -3 _____ 0

18. 19 _____ -38/2

19. -20 _____ -260

20. -567 _____ -350

9.3.2 Ordering integers.

Order the integers from the least to greatest.

1. 6, 0, -2, 4, 2 _____

2. -4, 5, 43, 6, -9 _____

3. 21, -11, -23, 4, 0 _____

4. 3, 5, -5, -44, 37, -36 _____

5. 2, -12, -1, -7, 17 _____

6. 32, 17, -16, -33 _____

7. 10, -10, 100, -100, -1000 _____

8. -99, -999, -9, 99, 0, -19 _____

9. -1, -2, -3 , 0, 1 ,2 ,3 _____

10. -66, -33, -44, 68,24, 32, -32 _____

11. 42, -32, 0, -43, -5, -33 _____

12. – 5, 80, 48, - 9, -8, 5, -15 _____

9.4. Add & Subtract Integers

9.4.1 Find the sum of the integers

1. (-7) + 2 = _____
2. (- 4) + 4 = _____
3. 3 + (-7) = _____
4. 10 + (-4) = _____
5. 15 + 41 = _____
6. (- 4) + 19 = _____
7. (- 9) + (-18) = _____
8. (-31) + 18 = _____
9. 71 + 89 = _____
10. (-5) + 5 = _____

11. (-66) + 0 = _____
12. (-10) + 19 = _____
13. (-30) + (-69) = _____
14. 12 + 23 + 43 = _____
15. (-12) + 6 + (-4) = _____
16. (-13) + 26 + (13) = _____
17. (-48) + (-16) + 50 = _____
18. (-10) + (-70) + (-20) = _____
19. 15 + (-37) + (-26) = _____
20. 8 + 123 + (-146) = _____

9.4.2 Find the difference

1. (- 5) – 0 = _____
2. 6 – 0 = _____
3. 4 – 2 = _____
4. 3 – 7 = _____
5. 5 – 4 = _____
6. (- 4) – 9 = _____
7. (-3) – 3 = _____
8. (-3) – (-3) = _____
9. 3 – (-3) = _____

10. (- 19) - (-18) = _____
11. (-31) – 18 = _____
12. 81 – 89 = _____
13. (-5) – 15 = _____
14. (-66) – 0 = _____
15. (-30) – 19 = _____
16. (-80) - (-69) = _____
17. (-32) – (-23) – 43 = _____
18. (-12) - 6 - (-4) = _____

19. (-22) - 44 - (-66) = _____

20. (-88) - (-16) – 50 = _____

21. (-100) - (-70) - (-20) = _____

22. 55 - (-37) - (-26) = _____

23. 8 - 123 - (-146) = _____

24. – 300 – (-423) – (-25) = ____

9.4.3. Multiply the integers

1. 1 × 1 = _____

2. 1 × 5 = _____

3. 3 × 3 = _____

4. -3 × 3 = _____

5. -3 × (-3) = _____

6. -13 × (-3) = _____

7. 4 × 2 = _____

8. 5 × 0 = _____

9. -6 × 5 = _____

10. -2 × -2 = _____

11. 7 × (-11) = _____

12. -8 × (-1) = _____

13. 9 × (-1) = _____

14. (-1) × (-1) = _____

15. (-4) × (-3) = _____

16. (-5) × (-3) × (-2) = _____

17. (-1) × (-10) ×12 = _____

18. 4 × 12 × 0 = _____

9.4.4. Divide the integers

1. - 2 ÷ 2 = _____

2. - 1 ÷ - 1 = _____

3. - 6 ÷ 2 = _____

4. 90 ÷ - 2 = _____

5. - 8 ÷ - 2 = _____

6. – 15 ÷ 3 = _____

7. -4 ÷ 2 = _____

8. 6 ÷ (-6) = _____

9. $0 \div 14 =$ _____

10. $(-8) \div (-4) =$ _____

11. $10 \div (-2) =$ _____

12. $1 \div 1 =$ _____

13. $14 \div 14 =$ _____

14. $(-15) \div 15 =$ _____

15. $(-15) \div (-15) =$ _____

16. $(-10) \div (-5) =$ _____

9.5. Coordinate Plane

Write the coordinates of each point labeled below

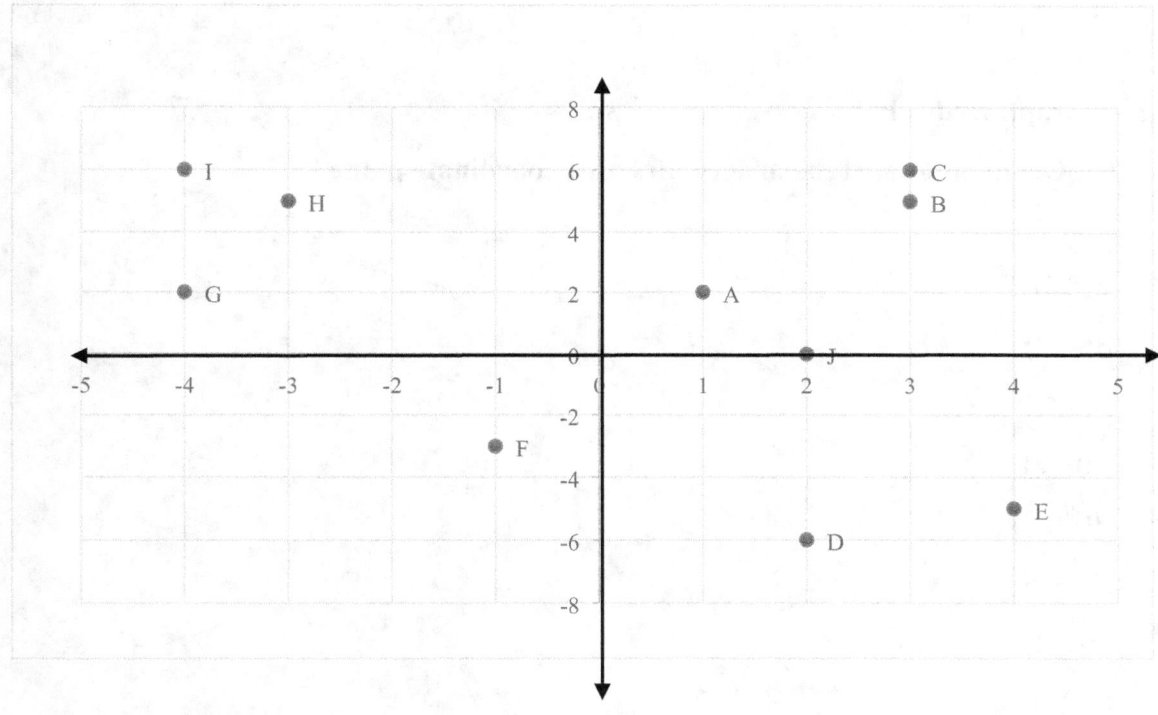

9.6. Quadrants and axis

Write the quadrant or axis that contains each point

1. $(4, -1) =$ _____

2. $(-2, -3) =$ _____

3. (- 6, 9) = _____

4. (15, 11) = _____

5. (20, - 30) = _____

6. (0, 5) = _____

7. (7, 0) = _____

8. (0,0) = _____

9. (-23, 8) = _____

10. (4, -5) = _____

9.7. Graph Order Pairs

9.7.1. Graph and label the order pairs on a coordinate plane

1. A(0, 0)
2. B(1, 2)
3. C(3, 2)
4. G(0, 5)
5. H(-2, 3)
6. D(-3, 6)
7. E(-5, 0)
8. F(-8, 1)
9. I(0, -4)
10. J(-3, -3)

9.7.2. Write the coordinates of the order pair described by the following Sentence.

1. Start from the origin. Move down two units and right 4 units. _____.

2. Start from the origin. Move left 3 units and up 5 units _____.

3. Start from origin. Move 4 units to the right and 4 units up _____

4. Start from origin. Move 5 units up and 3 units down _____.

5. Start at (2, 1). Move north 3 units and 2 units east _____.

6. Start at (-1, 3). Move south 2 units and 4 units west _____.

7. Start at (-6, -2). Move 6 units North and 2 units west _____.

8. Start at (4, -2). Move 5 units south and 1 unit south _____.

9. Start from origin. Move 10 units east and 5 units west _____.

10. Start from (3, -4). Move 3 units south, 2 units up, and 5 units west ____.

9.8. Function Rule & Graph

Looking at the pattern in the table, determine the missed values

1.

Input, x	1	2	3	4	5		6	7
Output, y	2	4	6	8	____		12	14

2.

Input, x	1	2	3	4	5	6	7
Output, y	3	6	9	__	15	18	__

3.

Input, x	-3	-2	-1	0	1	-	3
Output, y	-3	-2	-1	0	-	-	-

4.

Input, x	-3	-2	-1	-	1	3
Output, y	-4	-3	-2	-1	0	-

5.

Input, x	-3	-2	-1	0	1	2	3
Output, y	-	-	-4	0	4	8	12

6.

Input, x	-3	-2	-1	0	1	-	3
Output, y	-3	-2	-1	0	-	-	-

7.

Input, x	-3	-2	-1	0	1	-	3
Output, y	3	2	1	0	-1	-2	-3

8.

Input, x	-3	-2	-1	0	1	2	3
Output, y	6	4	2	-	-	-4	-

9.9 Map and Geometry on coordinate planes

Use the coordinate plane to answer the following map questions.

1. Towns A and B have the same y – coordinates. Town A is 5 units to the right of town B.

   a. A is at located 2 units up and 4 units to the right of the origin. Where is A located?

   b. Where is B located?

   c. If A was at (4, 1). Where would B be located?

   d. If A was 2 units to the left of the origin, where would B be located?

2. Bob's cousin lives 3 miles north and 4 miles east of Bob's house.

   a. If Bob's house is located at (0, 0), where would his cousin's house be located?

   b. If Bob's house is located at (-2, 1), where would his cousin's house be?

   c. If Bob's cousin wants to visit Bob at his house, he will go _____ units south and _____ units west.

   d. If Bob's cousin house is at (6, 8), where is Bob's house located? _____

3. Start at the point A (3, 2). Move 2 units up and 1 unit to the left to reach point B. Then Move 2 units down and 3 units east to get to point C. What are the coordinates for the point C?
   _____

4. Name the polygon that has vertices:

   a. (0, 0), (3, 0), and (0, 4) _____

   b. (1, 1), (3, 1), (3, 3), and (1, 3) _____

   c. (-2, 5), (4, 5), (4, 8), (-2, 8) _____

9.10. Word Problem on Integers

1. The temperature of a COVID –19 patient was 106° F. After taking medicine the patient's temperature was 7° lower. What is the patient's temperature?

2. A boy is on the fifth stair in the two-story house. He goes up 7 stairs and then down 4 stairs. At what stair is the boy?

3. A submarine deeps into the ocean 5 miles below the surface of the water, then goes up 2 miles. How far is the submarine from the surface of the water?

4. In winter the temperature drops so quickly. In a typical winter the temperature at night was 8° and dropped 13° within 3 hours. What is the temperature?

5. Bill borrowed $23 from his sister. His uncle gave him $50 for his birth day. He got $65 fine for parking violation. What is his balance at this point?

6. In Panama City it gets warmer in the afternoon than in the morning. There is a difference of 18° between 9 am and 3 pm. What is the temperature at 3 pm if the temperature at 9 am was -5°?

7. A glass of water has a temperature of 7° before putting it in the fridge. After 20 minutes, the temperature became -1°. What is the temperature difference?

8. The difference between temperatures of two cities is 71°. One of the cities has a temperature of -35°. What is the temperature of the other city?

9. According to the weather man, tomorrow's temperature at noon is as high as today's temperature. Today's temperature is 5° lower than 65°. What is tomorrow's temperature?

10. What is the difference between 120 and – 20?

11. A lumberjack climbs up 16 feet to cut the longest branch. He then goes down 6ft. to cut the next branch. What is the distance between the two branches?

12. What is the temperature difference between −12° C and 4° C?

13. A business man has the following transactions (gain and loss) in a week time: $1,500, -$600, $768, -$200, -$1200, $3000, $1200. Did the business man make money?

14. How many years are there between 240 BC and 2020 AD?

15. Team A scored the following points: 4, -3, 10, 6, and 10. Team B scored 6, -5, 11, -4, and 7. Which team is the winner?

16. Tom has $ 5, 000 in his bank. His credit card balance is $1234. If he pays off his credit card balance, how much money will be left in his bank account?

9.11. Chapter 9 Test

1. A figure has vertices (2, 1), (6, 1), (6, 5), and (2, 5). What figure will it be if you graph it on coordinate axes? _____

   a. Trapezoid
   b. Rectangle
   c. Square
   d. Triangle

2. What is the perimeter of the figure that has vertices: (1, 1), (6, 1), (6, 3), and (1, 3)

   a. 12
   b. 13
   c. 11
   d. 14

3. The opposite of 35 is = _____

   a. 0

   b. -35

   c. 35

   d. 3

4. The point (4, - 9) is in the _____ quadrant

   a. 1st

   b. 2nd

   c. 3rd

d. 4th

5. On a number line 4 is located _____
    a. To the right of 0
    b. To the left of 2
    c. To the right of 9
    d. To the left of a negative number
    e.

6. (-25) + 11 = _____

    a. 14

    b. -14

    c. – 36

    d. 36

7. 73 – (- 15) = _____

    a. 58

    b. 88

    c. -58

    d. -88

8. (-34) – 8 + 5 = _____

    a. - 37

    b. 37

    c. – 47

    d. – 21

9. (-28) – (-18) = _____

    a. 10

    b. – 46

    c. – 10

    d. 36

10. 9 × (- 5) = _____

    a. 45

b. -45

  c. 14

  d. -14

11. (- 8) × (-6) = _____

    a. -48

    b. 48

    c. 14

    d. -14

12. (-48) ÷ (- 12) = _____

    a. 4

    b. -4

    c. 6

    d. 12

13. Order the integer from greatest to smallest 42, -32, 0, -43, -5, and -33

    a. -5, 42, 0, -32, -33, -43

    b. 42, -5, 0, -32, -43, - 33

    c. 42, 0, -5, -32, -33, -43

    d. 0. 42, -43, - 33, - 5, - 32

14. Start from origin. Move 4 units left and 5 units down on the coordinate axes. Where is the point located?

    a. (4, 5)

    b. (-4, 5)

    c. (-4, -5)

    d. (0, 0)

15. If town A is located at (6, 8) and Town be is located at (5, 8). Then A and B lie on:

    a. The x – axis

    b. The y – axis

    c. Fourth quadrant

    d. First quadrant

16. The temperature at night was 8° and dropped 25° within 5 hours. What is the temperature?

   a. 13°

   b. -17°

   c. 33°

   d. -33°

17. Sally had $23. She sold candy and made $7. She gave $13 to her brother. How much money does she have now?

   a. $43

   b. $13

   c. $3

   d. $17

18. Looking at the pattern: 1, 3, 5, 7, 9, __. What is the next integer?

   a. 8

   b. 11

   c. 12

   d. 13

19. Mount Zion is 1200 ft. above sea level. The depth of Lake Tana is 20 ft. below sea level. How far is the vertical distance between Mount Zion and Lake Tana?

   a. 1400 ft.

   b. 100 ft.

   c. 1220 ft.

   d. 1180 ft.

20. In a coordinate plane, the point of intersection of the two axes is _____.

   a. X- coordinate

   b. Y – coordinate

   c. Origin

   d. (1, 3)

21. Consider the pattern in the table to determine missed values

Input, x	1	2	3	4	5	6	___
Output, y	5	10	15	20	___	30	35

    a. The missed x value is 7 and y value is 25

    b. The missed x value is 6 and y value is 25

    c. The missed x value is 7 and y value is 35

    d. The missed x value is 25 and y value is 7

22. $(-2 \times 8) - (3 \times (-4)) =$ _____

    a. 16

    b. 4

    c. -4

    d. -16

23. $-15 \times 10 \times 0 =$ _____

    a. -150

    b. 10

    c. -15

    d. 0

24. $12 - (-6) \div (-3) =$ _____

    a. 8

    b. -2

    c. 10

    d. -10

25. $(20 \div 4) - (0 \times (-4)) =$ _____

    a. 1

    b. -5

    c. 5

    d. 0

CHAPTER 10: RATIONAL NUMBERS

10.0 Short Notes on Real Numbers

- Rational number: integer together with fractions (repeating and terminating decimals)

- Repeating decimal example: 0.6666666…, 0.2323232323…

- Terminating decimal example: 1.25, 3.456

- Real number: a number that is rational or irrational

- Irrational number neither repeats nor terminates

- Absolute value of a number is the distance from zero to the number.

- Absolute value of a nonzero number is positive, $|-3| = 3$, $|5| = 5$, $|0| = 0$

- The distance between two numbers on a number line is the absolute value of the difference between the two numbers.

10.1 Terminating & Repeating decimal

Determine if each fraction or decimal is terminating or repeating

1. ¼ = _____

2. 2/3 = _____

3. 32/75 = _____

4. 1/10 = _____

5. 3/11 = _____

6. 6/10 = _____

7. 19/7 = _____

8. -1/3 = _____

9. 0.345 = _____

10. 3.333… = _____

11. 2.354 = _____

12. 14/18 = _____

13. 2/66 = _____

14. 4/13 = _____

15. 38/10 = _____

16. 10/9 = _____

17. 9/10 = _____

18. 34/44 = _____

19. 9/2 = _____

20. 2/9 = _____

10.2. Comparing Rational Numbers

10.2.1 Use < or > or = to compare the following rational numbers.

1. 1/3 _____ 1/2

2. 2/5 _____ 3/8

3. -5 _____ -15

4. ¾ _____ 0.75

5. π _____ 25/4

6. 0.666… _____ 0.68

7. 1.121314 … _____ 1.09

8. 1/3 _____ 2/7

9. 12 × 0 _____ 0× 11 + 2

10. 20/30 _____ 0.666 …

11. 7/10 _____ 0.07 × 10

12. 4 ÷ 100 _____ 2/100

13. -2/5 _____ 0.2

14. 0.001 _____ - 3/4

15. 0 _____ -23 × 5 × 0

16. 79 _____ 0.1 × 790

17. -4 × (-2) _____ 16

18. -0.001 _____ -1.6

19. -8.9 _____ 9.1 × (-10)

20. 2- (-4) _____ 7/2

10.2.2 Order the rational numbers from least to greatest.

1. -0.1, 2, 1/10, 2/5, 1/2

2. -6/10, -2/3, -1/2

3. -7/4, 1/10, -2/5, 5

4. -5, -5.01, -6, 0, 2

5. -7/11, -7/4, -1/10, -2/5, -13/12

6. 30/4, 14/5, 25/6, 64/7, 10/2

7. -3.75, -1.80, -2.833

8. -10.25, -10.4, -6.833, 10.4, 0.5

9. -3.20, -2.4, 15/6, 6

10. -18/7, 200/56, 215/36, -24.2

11. 16/5, 12/5, 15/16, 6/5, 10

12. 101/51, 104/51, 33/16, -16/5, 5

10.3. Absolute Value

Find the absolute value of each number.

1. $|-1/4| = $ _____

2. $|2.5| = $ _____

3. $|-1.43| = $ _____

4. $|-3/100| = $ _____

5. $|-3 + 12| = $ _____

6. $|-4 \times 4| = $ _____

7. $|-3^2| = $ _____

8. $|-10 \div 5| = $ _____

9. $|-2 - 3| = $ _____

10. $|-8 - (-11)| = $ _____

11. $|-3| = $ _____

12. $|4| = $ _____

13. $|0| = $ _____

14. $|82| = $ _____

15. $|-25| = $ _____

16. $|-5 + 6| = $ _____

17. | -2 + 4 | = _____

18. | 9 - 3 | = _____

19. | -3 × 4 | = _____

20. | 9 × 0 | = _____

21. | 1-(-5) | = _____

22. | -3- (-7) | = _____

23. | -3/4 | = _____

24. | -.645 | = _____

25. | 2.25 |= _____

26. | 7 - 2.5 | = _____

27. | 6 × (-2.5) | = _____

28. | 5 × 0.4 | = _____

29. | -2 × (-3) | = _____

30. | -4 × 3 × (-3) | = _____

31. | -11/4 | = _____

32. | -3.333 | = _____

33. | -π | = _____

34. | -2+7- (19) | = _____

35. | -8 + (-18) | = _____

36. | 27- (-15) | = _____

37. | -3.1111 + 1.3333 | = _____

38. | 2.444 - 1.333 | = _____

39. | 1.333 – 2.444 |= _____

40. | 2.444 – (-1. 333) | = _____

10.4 Distance on a number line and coordinate plane

Find the distance between the two numbers on a number line.

1. 3 and 8 = _____

2. – 4 and – 5 = _____

3. 3 and – 6 = _____

4. – 4 and – 8 = _____

5. 12 and 54 = _____

6. -12 and 88 = _____

7. – 4.5 and 7.5 = _____

8. 3.4 and – 12.8 = _____

9. 4.24 and –14. 76 = _____

10. – 0.3 and 11.7 = _____

Find the distance between the two points

11. (0, 0) and (0, 4) = _____

12. (0, 2) and (0, 7) = _____

13. (0, -3) and (0, 5) = _____

14. (1, 5) and (3, 5) = _____

15. (6, 3) and (-2, 3) = _____

16. (4, 5) and (-4, 5) = _____

17. (2, 3) and (7, 3) = _____

18. (1, 1) and (1, -5) = _____

19. (2.5, 3) and (7.5, 3) = _____

20. (-1/4, 2) and (¾, 2) = _____

10.5 Polygons on coordinate planes

Classify the polygon using the given vertices.

1. Vertices: (0, 0), (6, 0), and (0, 8)

2. Vertices: (1, 1), (6, 1), (1, 4), (4, 4)

3. Vertices: (2, 1), (5, 1), (7, 4), and (4, 4)

4. Vertices: (1, 0), (4, 0), (4, 3), (1, 3)

5. Vertices: (- 5, 1), (5, 1), (3, -3), (- 1, - 3)

Find the perimeter of each figure that has:

6. Vertices: (-1, 2), (3, 2), (3, 0), (-1, 0)

7. Vertices: (3, 1), (7, 1), (3, 5) (7, 5)

8. Vertices: (0, 0), (3, 0), (0, 4)

9. Vertices: (2, 0), (5, 0), (5, 3), (2, 3)

10. Vertices: (2, 1), (2, 5), (6, 5), (9, 2)

Find the area of each polygon determined by the given vertices.

11. Vertices: (0, 0), (6, 0), and (0, 8)

12. Vertices: (1, 1), (6, 1), (1, 4), (4, 4)

13. Vertices: (2, 1), (5, 1), (7, 4), and (4, 4)

14. Vertices: (1, 0), (4, 0), (4, 3), (1, 3)

15. Vertices: (- 5, 1), (5, 1), (3, -3), (- 1, - 3)

16. Vertices: (-1, 2), (3, 2), (3, 0), (- 1, 0)

17. Vertices: (3, 1), (7, 1), (3, 5) (7, 5)

18. Vertices: (0, 0), (3, 0), (0, 4)

19. Vertices: (2, 0), (5, 0), (5, 3), (2, 3)

20. Vertices: (2, 1), (2, 5), (6, 5), (9, 2)

10.6. Add & Subtract

Add or subtract the numbers.

1. $12 - 3.4 =$ _____
2. $(-5) + 6 =$ _____
3. $|(-4) - 3 - 7| =$ _____
4. $-4 - 6.25 =$ _____
5. $8.1 - (-3) =$ _____
6. $|(-15) + (-21)| =$ _____
7. $13.8 - (-2.2) =$ _____
8. $(-4.8) - 16.4 =$ _____
9. $|(-7) - (7/2)| =$ _____
10. $(-11.11) - (-0.11) =$ _____
11. $(13/2) - 2.5 =$ _____
12. $(-5) - 6 - 7 =$ _____
13. $13 + (-7) - 2 =$ _____
14. $(-17) - (-4) - 3 =$ _____
15. $(-5) - (-3) - (-6) =$ _____
16. $|12 - 6 + 4 - 7| =$ _____
17. $(-11) - (-5) - 6 + 2 =$ _____
18. $45 - (-9) - (-6) - 5 =$ _____
19. $|2.1 - 1.22 + 3.4| =$ _____
20. $(-13/2) - (-4.12) - (-2.3) =$ _____
21. $(-0.34) + 1.98 - (-2.45) =$ _____
22. $-4.5 - 3.61 - 2.91 =$ _____
23. $(-0.23) - (1.67) - (-2.5) =$ _____
24. $1 - (-0.01) - (-0.001) =$ _____
25. $(-1/2) - (-1/4) + 0.25 =$ _____
26. $5 - 0.214 - (-4.214) =$ _____

10.7. Multiply & Divide

10.7.1 Multiply each

1. $2 \times 15 =$ _____
2. $|4 \times (-4)| =$ _____
3. $(-6) \times 3 =$ _____
4. $|(-7) \times (-2)| =$ _____
5. $|(9 \times 2) \times (-4)| =$ _____
6. $(-5 \times -4) \times 3 =$ _____
7. $-4 \times -3 \times -6 =$ _____
8. $6 \times -11 \times -5 =$ _____
9. $11 \times -3 \times -5 =$ _____
10. $|-3 \times -4 \times -0| =$ _____
11. $20 \times -3 \times 41 =$ _____
12. $|-23 \times -2 \times 32| =$ _____
13. $|-21 \times -3 \times 13| =$ _____
14. $-11 \times -4 \times -18 =$ _____
15. $-32 \times 12 \times -12 =$ _____
16. $-42 \times -11 \times -100 =$ _____
17. $-0.01 \times 100 \times -10 =$ _____
18. $-0.02 \times 0.11 \times 10000 =$ _____

19. -0.12 × -200 × 10 = _____

20. -0.12 × -200 × -10 = _____

21. – (-56) × -800 = _____

22. 1.6 × -0.12 × 1000 = _____

23. 11 × 11 × -11 = _____

24. 0.01 × 0.01 × 10000 = _____

10.7.2 Divide each

1. 10 ÷ 2 = _____
2. |9 ÷ (-3)| = _____
3. -18 ÷ 6 = _____
4. -24 ÷ 12 = _____
5. 180 ÷ 6 = _____
6. |-240 ÷ (-8)| = _____
7. |-360 ÷ 12| = _____
8. 480 ÷ -24 = _____
9. (-46/5) ÷ (-25) = _____
10. – (- 18) ÷ (2/3) = _____
11. (-150/5) ÷ 6 = _____
12. -210 ÷ (1/3) = _____
13. (4/5) ÷ (-4/25) = _____
14. – (7/3) ÷ (-21/9) = _____
15. 21 ÷ (7/3) = _____
16. 480 ÷ -24 ÷ 20 = _____
17. -1 ÷ -24 ÷ (1/24) = _____
18. (-1/3) ÷ (1/3) = _____
19. -1 ÷ (-1/10) ÷ (-1/10) = _____
20. 0 ÷ 100 ÷ 0.383474 = _____

10.8. Word Problems on Real Numbers

1. The sum of two numbers is 3.46. One of the numbers is 2. What is the other number?

2. John paid $10.00 to the casher for a sandwich and a drink. The cost of the sandwich was $4.56 and of the drink was $1.25. How much change did he receive?

3. Lora planned to score 95 in her test. She scored 4.5 less than what she had planned. What is her actual score?

4. In the second test (refer question 3), She scored 3.9 more than what she had planned. What is her second test score?

5. Max ran 2.14 miles on Monday. On Tuesday he ran twice as much as Monday. Altogether how many miles did Max run?

6. A builder is placing bricks one on the top of the other. The height of the bricks was 18 ft. how many bricks are there if each brick is 0.5 ft. high?

7. The distance between two numbers on a number line is 6. If one of the numbers is 3 and the other number is a negative integer. What is the other number?

8. In question 7, if the other number was a positive integer, what would the number be?

9. The absolute value of two different numbers is 7, what are the numbers?

10. The length of the side of a regular hexagon is 2.34 in. What is the perimeter?

10.9. Chapter 10 Test

1. Find the length between the points A (4, 6) and B (- 2, 6).

   a. 2
   b. 5
   c. 7
   d. 6

2. Use the vertices A (5, 2), B (5, 4), C (2, 4), and D (2, 2) to classify the figure.

   a. Trapezoid
   b. Rectangle
   c. Square
   d. Rhombus

3. Find the perimeter of the polygon that has vertices: A (1, 1), B (1, 11), C (7, 11), and D (7, 1)

a. 32
b. 16
c. 60
d. 40

4. Find the area of a triangle that has vertices A (0, 0), B (6, 0), and C (0, 8).

   a. 48
   b. 24
   c. 36
   d. 44

5. What is the distance between 5/4 and – 3/4?

   a. ½
   b. ¼
   c. 2
   d. 1

6. The point (5, - 4) is in the _____ quadrant.

   a. 1st
   b. 2nd
   c. 3rd
   d. 4th

7. The point (- 3, - 6) is in the _____ quadrant.

   a. 3rd
   b. 2nd
   c. 1st
   d. 4th

8. Write 10/24 as a decimal.

   a. 0.4
   b. 0.41666…
   c. 0.5
   d. 1.2

9. – 1/3 is a _____

   a. Terminating decimal
   b. Repeating decimal
   c. Irrational number
   d. Integer

10. The fraction 1/8 is a _____

    a. Terminating decimal

b. Repeating decimal
c. Negative integer
d. Irrational number

11. The correct sign that makes – 5 _____ |- 10| true is_____.

    a. <

    b. >

    c. =

    d. No answer

12. 6 _____ 18 ÷ 3

    a. =

    b. >

    c. <

    d. No answer

13. Order -5, 7, -0.555, 4, 6.55, and | -10 | from least to greatest

    a. | -10|, 7, -0.555, 4, -5, 6.55

    b. -5, -0.555, 4, 6.55, 7, |- 10|

    c. 7, -5, -0.555, 4, 6.55, | -10|

    d. -0.555, 4, 6.55, -5, | -10|, 7

14. Order from greatest to least: -6, 3, -10, -8, and |-2|

    a. – 10, - 8, - 6, 3, |-2|

    b. – 10, - 8, - 6, |-2|, 3

    c. – 6, - 10, - 8, 3, |-2|

    d. 3,|-2|, -6, -8, - 10

15. | - 3| = ____

    a. 5

    b. -3

      c. 3

      d. None

16. 4 – (- 18) + 5 = _____

      a. 27

      b. 9

      c. – 18

      d. 36

17. -25 – (33) – 2 × (-5) = _____

      a. 48

      b. – 48

      c. 38

      d. 58

18. |-8 – (- 10) – 5| = _____

      a. 25

      b. -3

      c. 3

      d. 6

19. |-5 – 3 - 2× (-5) | = _____

      a. -2

      b. 18

      c. -18

      d. 2

20. 3 × (-5) = _____

      a. 15

      b. |-15|

      c. -15

      d. 8

21. 45 – (-3 × 4) = _____

    a. 48

    b. 192

    c. 33

    d. 57

22. -12 – (-6 ÷ 1/2) = _____

    a. -24

    b. 24

    c. 0

    d. -9

23. |- 60 ÷ 2 ÷ (-3) | = _____

    a. 10

    b. -10

    c. 20

    d. 360

24. |6 – (-3 ÷ 1/3) |

    a. 7

    b. 15

    c. 5

    d. -8

25. -.001 × 10000 – (- 5 × 10)

    a. 50

    b. 60

    c. 70

    d. 40

26. Mike runs 5 miles to north and then 4 miles west starting from the origin. Where is Mike now?

    a. (4, 5)

    b. (5, 4)

c. (-4, 5)

d. (-4, -5)

27. Where will be (-3, 4) if the point moved 3 units to the left and 4 units down?

    a. (0, 0)

    b. (6, 8)

    c. (-6, 0)

    d. (8, -6)

28. $|2000 \div 100 - (-5 \times 10)| = $ _____

    a. 30

    b. 70

    c. 40

    d. 100

29. The absolute value of the difference between -14.5 and -22.5 is _____.

    a. -8

    b. 12

    c. 8

    d. -12

30. The perimeter of a trapezoid with side lengths 3.14cm, 2.54cm, 5cm, and 6cm is:

    a. 16.68 cm

    b. 16.8 cm

    c. 14.68 cm

    d. 28 cm

CHAPTER 11: PRACTICE EXAMS

Standardized/ End of year Practice Exam 1

1. Estimate the sum to the nearest whole number: 2.36 + 5.8

    a. 9

    b. 8.16

    c. 8

    d. 9

2. Convert 3.5 miles to yards.

    a. 5,280 yards.

    b. 15, 840 yards.

    c. 15,000 yards.

    d. 6160 yards.

3. Reduce 24/72 to the simplest fraction

    a. 3/9

    b. 3/72

    c. 1/3

    d. 3

4. To evaluate the numerical expression. $\{75 - 12 \div [(9 \times 4) + 0]\}$

What operation should be performed first?

    a. Divide 12 by 9.

    b. Multiply 9 by 4.

    c. Subtract 75 from 52.

    d. Add 36 and 0.

5. $11\frac{3}{4} - 9\frac{3}{4} =$ ___

    a. 3

    b. 2

    c. -2

    d. No answer

6. Which of the following is equivalent to 4/7

    a. $\frac{30}{70}$

    b. $\frac{9}{21}$

    c. $\frac{18}{79}$

    d. $\frac{28}{49}$

7. Estimate 71% of 804.

    a. 700

    b. 560

    c. 800

    d. 670

8. $3x + 21y =$ _____

    a. $2(x + 7y)$

    b. $3x(7y + 1)$

c. 3(x + 7y)

d. 3x

9. The sequence 1, 4, 16, 64 … is:

   a. Arithmetic sequence

   b. Geometric sequence

   c. Neither arithmetic nor geometric

10. Put the following decimals in order from least to greatest. 0.5218, 0.5812, 0.5821, & 0.5182

   a. 0.5218, 0.5812, 0.5821, 0.5182

   b. 0.5182, 0.5218, 0.5812, 0.5821

   c. 0.5182, 0.5812, 0.5218, 0.5821

   d. 0.5812, 0.5218, 058, 0.5182

11. The solution of: 3x < 21 is:

   a. X < 7

   b. X > 7

   c. X < 21

   d. X > 21

12. The opposite of the number – 456 is

   a. 0

   b. 456

   c. 1/456

   d. – 1/456

13. If you run 3 miles an hour, how long will it take you to run 3.6 miles?

   a. 1. 2 hours

   b. 12 hours

   c. 5 hours

   d. 6.5 hours

14. In a car dealer shop, 40 % are new cars. Write this as decimal

   b. 44

   c. 4

   d. 0.04

   e. 0.4

15. In question 14, if there are 400 cars, how many cars are old?

   a. 160

   b. 280

   c. 240

   d. 300

16. Bob runs 100 miles in 20 hours. What is his average speed?

   a. 5 mph (miles per hour)

   b. 9 mph

   c. 10 mph

   d. 15 mph

17. Write the inequality for the statement. A number is at least 17.

    a. x > 17

    b. x < 17

    c. x ≤ 17

    d. x ≥ 17

18. The third quartile of the numbers: 2, 5, 4, 9, 7, 6, and 8 is:

    a. 9

    b. 8

    c. 7

    d. 6

19. The point A (3, -4) is in the _____ quadrant

    a. 4th

    b. 1st

    c. 3rd

    d. 2nd

20. 1/7 is a:

    a. Repeating decimal

    b. Terminating decimal

    c. integer

    d. Irrational number

21. What is the median of the following numbers?

300, 39, 71, 39, 86, 38, 25

   a. 42

   b. 39

   c. 38

   d. 35.5

22. What number would you divide by to calculate the mean of 18, 7, 1, 3, 4, 5, and 6?

   a. 8

   b. 4

   c. 7

   d. 4

23. What measure of center can have more than one value?

   a. Median

   b. Mean

   c. Mode

   d. Typical value

24. Given the numbers: 1, 2, 1, 9, 5, 3, and 20

   a. 20 is an outlier

   b. 9 is the mode

   c. 1 is the median

   d. 20 is not an outlier

25. The distance between -11.3 and 5.7 on a number line is:

   a. 5.6

   b. 4.8

   c. 17

   d. 6

26. The distance between A (3, 4) and (8, 4) is:

   a. 5

   b. 6

   c. 7

   d. 8

27. A table that has classes and corresponding frequencies is:

   a. A frequency distribution

   b. A histogram

   c. A pie chart

   d. A bar graph

28. The three angles of a triangle are 20°, 90°, and 70°. Based on the three angles, the triangle is a _____.

   a. Scalene triangle

   b. Equilateral triangle

   c. Right triangle

   d. Isosceles triangle

29. |-45| - 10 = _____

   a. 55
   b. - 35
   c. 35
   d. 25

30. -93 – (-25)

   a. -68
   b. 58
   c. 48
   d. -88

31. (-34) – (-18) × 2

   a. - 2
   b. 2
   c. 70
   d. - 70

32. 38 – (-18)

   a. 10
   b. – 46
   c. – 10
   d. 56

33. The correct sign that makes $-17 + |-16|$ _____ 0 true is.

   a. <

   b. >

   c. =

   d. No answer

34. The LCM of the numbers 40 and 15 is:

   a. 30

   b. 60

   c. 120

   d. 45

35. Order -85, 7, -0.555, 4, 6.55, | -10 | from least to greatest

   a. | -10|, 7, -0.555, 4, -5, 6.55

   b. -85, -0.555, 4, 6.55, 7, |- 10|

   c. 7, -5, -0.555, 4, 6.55, | -10|

   d. -0.555, 4, 6.55, -5, | -10|, 7

36. Which of the following numbers satisfies the inequality: $2x + 5 < 12$

   a. 5

   b. 6

   c. 4

   d. 0

37. The expression: $3x + 2$ is read as.

   a. Three times a number plus two.

   b. Two times a number plus three.

   c. Three times the product of a number and two.

   d. Three times a number.

38. Which of the following is the solution to the equation $2x - 4 = 8$

a. 5

   b. 4

   c. 8

   d. 6

39. To make box and whisker plot, which is not necessary

   a. minimum value

   b. maximum value

   c. Mode

   d. Median

40. Name the property used in the equation: $6(7+5) = 6 \times 7 + 6 \times 5$.

   a. Distributive property of multiplication over addition.

   b. Commutative property of multiplication.

   c. Associative property of multiplication.

   d. Commutative property of addition.

Final Exam practice 2

Standardized/ End of year Practice Exam

1. Estimate the difference: $9.9 - 4.1$

   a. 5.7

   b. 5

   c. 6

   d. 4

2. 10 is 40% of what number?

   a. 40

   b. 30

   c. 25

   d. 20

3. Evaluate: 0.1 × {2[8+2(11 - 9) −2]}

   a. 5

   b. 4

   c. 3

   d. 2

4. The surface area of a cube is 150 square meters. The length of each side = _____

   a. 5 meters

   b. 10 meters

   c. 4 meters

   d. 25 meters

5. Change 10/15 to percent

   a. 33%

   b. 66.7%

   c.   99%

   d. 12%

6. The GCF of 12 and 68 is:

   a. 5

   b. 17

c. 3

d. 4

7. The prime factorization of 204 is

    a. 4 × 68

    b. 2 × 2 x 17

    c. 2 × 2 × 3 × 17

    d. 17

8. Divide: $\dfrac{16}{72} \div \dfrac{2}{9}$

    a. 9

    b. 1

    c. $\dfrac{2}{3}$

    d. $\dfrac{2}{27}$

9. Find the next term of the arithmetic sequence: 1, 7, 13, 19, 25, ____.

    a. 27

    b. 28

    c. 30

    d. 31

10. Multiply 7.188 x 0.01 × 1000

    a. 71.88

b. 7.188

c. 0.7188

d. 718.8

11. Three times a number is added to 9 to get 36. What is the number?

    a. 8

    b. 9

    c. 10

    d. 12

12. Estimate: 89.4 ÷ 5

    a. 17.88

    b. 18

    c. 19

    d. 20

13. The area of a trapezoid with bases 6cm. and 8cm. and height 10cm. is _____.

    a. 35cm^2

    b. 140 cm^2

    c. 70 cm^2

    d. 144 cm^2

14. Which is not equivalent to 6/16

    a. 24/9

b. 3/8

c. 9/24

d. 12/32

15. Jasmine can buy 10 kiwi fruit for $2.5. How many kiwi can she buy for $7.5?

    a. 40

    b. 30

    c. 10

    d. 100

16. 10560ft. = _____

    a. 2 miles

    b. 3 miles

    c. 4 miles

    d. 5 miles

17. 40% of a school's sales come from selling books. If the school sold $2,000 worth of books. What was the school's total sales?

    a. $500

    b. $5,000

    c. $1,000

    d. $10,000

18. What percent of 60 is 21?

    a. 35%

b. 40%

c. 45%

d. 50%

19. The Interquartile range of the following set of numbers is:

50, 51, 85, 69, 19, 60, 80, 64, 109, 63

   a. 39

   b. 38

   c. 19

   d. 29

20. The mean weight of five doors is 167.2 pounds. The weights of four of the doors are 158.4 pounds, 162.8 pounds, 165 pounds, and 178.2 pounds, respectively. What is the weight of the fifth door? _____

   a. 171.6

   b. 166.6

   c. 157.4

   d. 177.6

21. A _____ is quadrilateral with four equal sides

   a. Rhombus

   b. Trapezoid

   c. Rectangle

   d. Kite

22. The longer side is 8m and the shorter side is 4m. Calculate the area of the polygon.

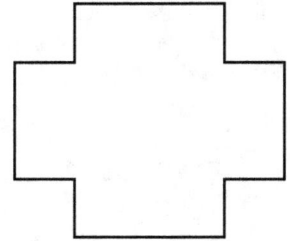

a. 128m²

b. 64m²

c. 192m²

d. 190m²

23. Find the surface area of a rectangular prism with length 5 in. width 6in. and height 7in.

a. 120 in²

b. 214in²

c. 375 in²

d. 50 in²

24. The solution of 0.25x = 4 is:

a. 16

b. 20

c. 15

d. 14

25. Identify an expression not equivalent to 4x + 6y

   a. 2x + 6y + 2x

   b. x + 2y + 4y + 3x

   c. 2(2x + 6x)

   d. 2(2x + 3x)

26. (- 30) × (-6.5) = _____

   a. -180

   b. 195

   c. 140

   d. 240

27. (-848) ÷ (212)

   a. - 4

   b. 4

   c. 6

   d. 12

28. What is the value of the expression $3y^3 + 4x^2$ when x = 0.5 and y = 2?

   a. 20

   b. 24

   c. 25

d. 2

29. The inequality for the graph below is:

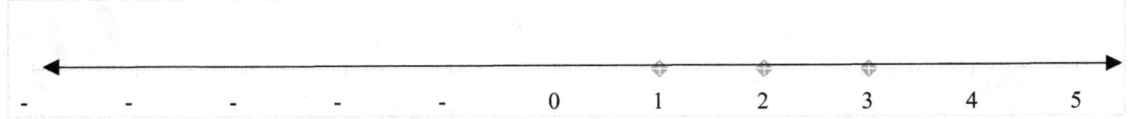

   a. x > 0

   b. x ≥ -1

   c. x < 2

   d. x > 5

30. 20 – (- 18) + 5 = _____

   a. 27

   b. 9

   c. – 18

   d. 43

31. -43 – 33 - |4× (-4) | = _____

   a. - 40

   b. – 18

   c. - 92

   d. 40

32. |-15 – (- 10) – 15| = _____

   a. 20

b. -3

c. 3

d. 6

33. Name the property used in the equation: $(9 \times 6) \times 11 = 9 \times (6 \times 11)$.

   a. Commutative property of addition

   b. Commutative property of multiplication

   c. Associative property of multiplication

   d. Distributive property of addition

34. Solve the equation: $2x - 54 = 2$.

   a. 10

   b. 34

   c. 16

   d. 28

35. The polygon that has vertices (0, 0), (4, 0), (4, 4), and (0, 4) is

   a. Rectangle

   b. Square

   c. Trapezoid

   d. kite

36. Another form of writing: $y \times (3 + 8)$ is_____.

   a. 3y + 8y

b. 12y

c. 3y + 8

d. 3 + 8y

37. Solve each equation: -4x – 8 = - 16

   a. 3

   b. 2

   c. 1

   d. 1.5

38. The point A (x, y) is in the 2nd quadrant. Which the following is true about A.

   a. x is positive and y is positive

   b. x is positive and y is negative

   c. x is negative and y is negative

   d. x is negative and y is positive

39. What is the probability of a 2 and a 5 showing up in rolling a die?

   a. 2/6

   b. 0

   c. 2/3

   d. 5/6

40. Solve - 0.5 x – 8 > 1.5x - 10

   a. x = -1

b. x = 1

c. x > 1

d. x < 1

Final Exam practice 3

Standardized/ End of year Practice Exam 3

1. [(5+3) - (1+3)] × (1/4) – 4 = _____

   a. 13

   b. 6

   c. 5

   d. - 3

2. Jasmine earned $450 working for n, number of weeks. Write an algebraic expression for the amount she earned each week

   a. 450/n

   b. 450n

   c. 450 + n

   d. 450 - n

3. If the product of 5 and a number is increased by 8, the result will be 58. What is the number?

   a. 16

   b. 22

   c. 10

   d. 5

4. Evaluate: $3x^2 - y^3$ for $x = 2$ and $y = -1$

   a. 11
   b. 13
   c. 10
   d. 12

5. Divide $2\dfrac{5}{6} \div 3\dfrac{3}{7}$

   a. $\dfrac{16}{42}$
   b. $\dfrac{68}{7}$
   c. $\dfrac{408}{40}$
   d. $\dfrac{119}{144}$

6. The order of the numbers: 2/3, 6/10, 1/2 is _____.

   a. Least to greatest
   b. Greatest to least
   c. There is no order

7. Thomas earns $7.4 an hour. How many hours should he work to make $66.6?

   a. 6 hours
   b. 7 hours
   c. 8 hours
   d. 9 hours

8. The area of a rectangular playground is 936 square meters and its length is 46.8 meters. Find the perimeter of the playground?

   a. 133.6 meters

   b. 187.2 meters

   c. 400 meters

   d. 21.6 meters

9. The 1st quartile of the numbers: 3, 9, 2, 6, and 5 is:

   a. 3

   b. 2

   c. 2.5

   d. 5

10. What is the estimated sum to the nearest tenth 8.61 - 4.09

    a. 0.8

    b. 4.5

    c. 4.52

    d. 5.6

11. What is 0.38941 × 2.4 rounded to 3 decimal places?

    a. 0.93

    b. 0.935

    c. 0.934

    d. 0.999

12. Which of the ordered pairs is a solution to the equation: $y = 3x - 1$

a. (-1, 2)

b. (2, 5)

c. (2, 4)

d. (0, 0)

13. What is the base of a parallelogram with an area of 33 square feet and a height of 60 feet?

    a. 0.55ft.

    b. 5.5ft.

    c. 6 ft.

    d. 5ft.

14. Which of the following is an example of terminating decimal?

    a. 1/3

    b. 2/3

    c. 1/8

    d. 0.666…

15. The chance to pass a test is 81.6%. What is the chance to fail in decimal form?

    a. 0.81

    b. 816/1000

    c. 0.816

    d. 0.184

16. Which of the following describes how to convert 34 yards into inches?

    a. Multiply by 12

b. Multiply by 36

c. Divide by 36

d. Divide by 3

17. Find the area of a triangle that has vertices: (0, 0), (0, 5), and (12, 0)

    a. 30 square units

    b. 40 square units

    c. 50 square units

    d. 60 square units

18. Dave walk with his dog 0.14 km around a neighbor. How many centimeters did he walk?

    a. 1,400 cm

    b. 14,000 cm

    c. 140 cm

    d. 4,400 cm

19. The point A (0, 8) is:

    a. On the y-axis

    b. In the first quadrant

    c. In the second quadrant

    d. In the third quadrant

20. What is the median of the following numbers? 10, 38, 71, 42, 38, 76, 37, 25

a. 42.5

b. 39

c. 38

d. 35.5

21. The following set of data represents the temperatures at different times of the day; 58, 79, 81, 99, 68, 92, 76, 84, 53, 57, 81, 91, 77, 50, 65, 57, 51, 72, 84, 89. If you want to make a frequency table, what will be the frequency of the interval 50 – 60.

a. 3

b. 4

c. 5

d. 6

22. What is the range of the numbers? 68, 55, 70, 62, 71, 58, 81, 82, 63, 79

a. 68

b. 27

c. 69

d. 79

23. The volume of a cube is 27,000 m³, its surface area = _____

a. 90 m

b. 10 m²

c. 100 m²

d. 5.400 m²

24. The volume of a rectangular prism with length 20in., width 5in., and height of 10in. is

   a. 60 in³
   b. 1000 in³
   c. 960 in³
   d. 80 in³

25. The volume of square prism is 150 ft² and the height is 6 ft. What is the side length of the base of the prism?

   a. 25 ft.
   b. 50 ft.
   c. 5 ft.
   d. 30 ft

26. Find the radius of a circle whose diameter is 10 cm?

   a. 5 cm
   b. 10π cm
   c. 10 cm
   d. 10 cm²

27. Start from (1, 1). Move 4 units left and 5 units up on the coordinate axes. Where is the point located?

   a. (4, 5)
   b. (-4, 5)
   c. (-4, -5)
   d. (-3, 6)

28. If town A is located at (7, 0) and Town be is located at (18, 0). Then A and B lie on:

   a. The x – axis
   b. The y – axis
   c. First and second quadrant
   d. Fourth quadrant

29. The temperature at night was 8° and dropped 12° within 3 hours. What is the temperature?

   a. 13°
   b. - 14°
   c. - 4°
   d. - 33°

30. Sally had $53. She sold candy and made $17. She gave $15 to her brother. How much money does she have now?

   a. $43
   b. $55
   c. $3
   d. $17

31. Your grades on math tests are 78, 64, 90, 85, 88. What would your next grade need to be to bring your average up to an 84?

   a. 89
   b. 95

c. 97

d. 99

32. |8 × (-5)| - (-15) = _____

   a. 25

   b. 30|

   c. 55

   d. 8

33. Solve the inequality: 3x > 3

   a. x < 3

   b. x > 3

   c. x < 1

   d. x > 1

34. Write an inequality for the sentence: You must be over 5 years old to go to elementary school

   a. x < 6

   b. x > 5

   c. x > 6

   d. x < 8

35. Solve each equation: -2x - 6 = 10

   a. - 8

   b. 8

   c. 0

d. -10

36. Solve: 3x − 4 = 2x - 6

   a. 2

   b. - 2

   c. 4

   d. -1

37. Find a number satisfying 2x + 4 = - 4 + 24

   a. 0

   b. -1

   c. No number satisfies the equation

   d. 3

38. Solve the equation: 0.8x = 48

   a. 50

   b. 60

   c. 80

   d. 70

39. Which one of the following can be probability of passing a test?

   a. 11/9

   b. - 3/9

   c. 4/9

   d. 50/9

40. How many numbers satisfy the inequality | x + 4| > 0

   a. infinite numbers

   b. 2 numbers

   c. 10 numbers

   d. No number satisfy the inequality

Final Exam practice 4

Standardized/ Milestone/ End of year Practice Exam 4

1. Estimate 19.7 ÷ 4.01 = _____

   a. 5

   b. 4

   c.

   d. 4.9127

2. In the numerical expression, 90 × (7 - 5 + 4) which should be calculated first?

   a. -5 + 4

   b. 90 × 7

   c. 7 - 5

   d. 7 + 5

3. Write the fraction, 420/ 1050 in simplest form

   a. 3/5

   b. 2/5

   c. 1/5

d. 4/5

4. Compare: (8 ÷ 2) × 3 _____ - (- | 6 – 7|) + 12)

   a. Cannot be compared.
   b. =
   c. >
   d. <

5. Evaluate y = 1 - 3x if x = -3

   a. Y = 10
   b. Y = - 8
   c. Y = 9
   d. Y = - 9

6. What is 86% 250?

   a. 200
   b. 215
   c. 205
   d. 255

7. Group A has a set of numbers: 1, 2, 3, 3.1, and 4. Group B has a set of numbers: 2, 3, 1, 25, and 2000. What measure of center best fits for group A?

   a. Mode
   b. Median
   c. Range
   d. Mean

8. In question 7, the best measure of center for group B is

   a. Mean
   b. Median
   c. Mode
   d. range

9. In question 7, what is best measure of variation (spread) for group A

   a. Mean absolute deviation
   b. range
   c. Interquartile range
   d. median

10. In question 7. Which group is more variable?

    a. Group A
    b. Both are the same
    c. Group B
    d. We cannot compare the groups

11. Divide 32.87 by 9 and round the quotient to the nearest hundredths

    a. 3.65
    b. 3.652
    c. 3.7

d. 3.66

12. 8.4 ÷ 2.1 ÷ 0.5 is

   a. 4
   b. 2
   c. 8
   d. 1

13. Name the quadrant that contains the point (-7, 9)

   a. 4th
   b. 3rd
   c. 1st
   d. 2nd

14. Find the missing number 5, 22, 2, __ when the mean is 13

   a. 24
   b. 23
   c. 25
   d. 26

15. 75 % of students in a school are boys. How many students are girls if the school has 2000 students?

   a. 600
   b. 400
   c. 500

d. 1000

16. The opposite of the number 1.01

    a. 1/1.01

    b. 1.11

    c. − 1.01

    d. 0

17. Which is equivalent to 100 grams?

    a. 0.1kg

    b. 100 milligrams

    c. 1kg

    d. 100 kg

18. If you toss three coins at a time, how many different outcomes do you expect?

    a. 4

    b. 8

    c. 6

    d. 10

19. Which is less: 3 gallons or 9 liters?

    a. 9 liters

    b. 3 gallons

    c. both are equal

20. The height of a new baby born was 16 inches. In three months, his height is 8 inches more. What is the height of the baby after three months?

   a. 25 inches

   b. 2 ft.

   c. 9 inches

   d. 32 inches

21. What is the distance between the two points A (0, 6) and B (0, - 4) on a coordinate plane?

   a. 11

   b. 9

   c. 11

   d. 10

22. Determine the polygon that has vertices: (1, 1), (3, 1), (3, 6), and (1, 6)

   a. Square

   b. Trapezoid

   c. Triangle

   d. Rectangle

23. The mean deviation of the numbers 1, 2, 3, 4, and 5 is:

   a. 2

   b. 1

   c. 1.2

   d. 3

24. What is the radius of a circle if the area is 9π or 28.26 square units?

   a. 1 unit

   b. 2 units

   c. 3 units

   d. Unknown

25. One of the solutions of: $2x - 4 < 20$ is:

   a. 15

   b. 20

   c. 5

   d. 12

26. The surface area of a box with length 6m, width 2m and height of 4m is:

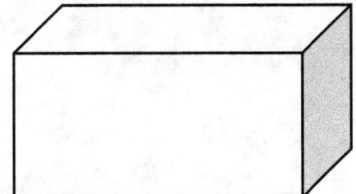

a. 88 m²

b. 44 m²

c. 56 m²

d. 76 m²

27. The area of a rectangular guest house is 4,800 ft². Each guest stays in a 10ft. by 10 ft. room. How many guests can the guest house accommodate?

  a. 100

  b. 48

  c. 80

  d. 40

28. Looking at the pattern: 1, 7, 14, 21, 28, __. What is the next integer?

  a. 32

  b. 40

  c. 42

  d. 35

29. Mount Zion is 800 ft. above sea level. The depth of Congo River is 700 ft. below sea level. How far is the vertical distance between mount Zion and Congo River?

  a. 2400 ft.

  b. 100 ft.

  c. 1500 ft.

  d. 1180 ft.

30. In a coordinate plane, x and y coordinates are negative in _____ quadrant.

  a. 3rd

  b. 2nd

  c. 1st

  d. 4th

31. looking at the pattern in the table, determine the missed values

Input, x	1	2	3	4	5	6	-
Output, y	6	12	18	24	___	36	42

   a. The missed x value is 7 and y value is 25
   b. The missed x value is 7 and y value is 30
   c. The missed x value is 7 and y value is 35
   d. The missed x value is 25 and y value is 7

32. The solution of: - 5x + 6 > 31

   a. x > 5
   b. x < 5
   c. x > - 5
   d. x < - 5

33. |6 ÷ (-3 ×1/3) | - 6

   a. 3
   b. 12
   c. 0
   d. -8

34. (-11) – (-5) – 6 + 8 = _____

   a. - 5
   b. - 6
   c. - 7
   d. - 4

35. John runs 7 miles to west and then 3 miles south starting from the origin. Where is Mike located now?

   a. (6, 7)
   b. (- 7, - 3)

c. (0, - 3)

d. (-4, -5)

36 The equation is true $3x + 6 = 24 + 2x$, for x = ___.

    a. 18

    b. 80

    c. 19

    d. – 9

37. $-2x - 5 = 5 - 2x$ is true for x = _____

    a. 3

    b. No number satisfies it

    c. -5

    d. all numbers satisfy the equation

38. $-10x + 17 = 17$ is true for x = _____

    a. -2

    b. 1

    c. 0

    d. 5

39. Use the formula: $A = \pi r^2 h$ to find h when $A = 28\pi$ and $r = 2$

    a. 6

    b. 3

    c. 4

    d. 7

40. Which value is considered as an outlier?

a. The first quartile

b the median

c. a number greater than the sum of the third quartile and twice the interquartile range.

d. The mode.

Final Exam practice 5

Standardized/ End of year Practice Exam 5

1. Looking at the pattern: 3, 5, 8, 13, 21, 34, ___, what is the missed number?

    a. 65

    b. 68

    c. 55

    d. 70

2. If x = 5, what is the value of the expression $(2x - 1)^2$?

    a. 101

    b. 99

    c. 121

    d. 81

3. Tom cut long ruler into 16 pieces; each piece is 5/4 inches. What was the original length of the ruler before Tom cut the ruler?

    a. 20 inches

    b. 16 inches

    c. 9 inches

d. 10 inches

4. Mr. Alex wants to cut a 10 feet rope that are ¼ ft. long. How many pieces can she get from this rope?

   a. 30

   b. 28

   c. 40

   d. 44

5. $2\frac{5}{6} - 3\frac{3}{6} = $ _____

   a. $\frac{16}{42}$

   b. $-\frac{2}{3}$

   c. $\frac{408}{40}$

   d. $\frac{119}{144}$

6. Order from least to greatest the fractions 1/6 3/5, 7/6, 1/3, 4/9.

   a. 1/6, 1/3, 4/9, 3/5, 7/6

   b. 1/6, 4/9, 1/3, 3/5, 7/6

   c. 1/6, 1/3, 4/9, 7/6, 3/5

   d. 1/6, 1/3, 3/5, 4/9, 7/6

7. Nancy and her family had lunch at their favorite restaurant. The total bill was $63.00, and they wanted to leave an 18% tip. Which amount of money is closest to the 18% tip?

a. $11

b. $12

c. $13

d. $14

8. What percent of 600 is 270?

   a. 25%

   b. 26%

   c. 27%

   d. 45%

9. Which choice gives the correct order of operations to evaluate the expression below?

   2 − (40 + 3) × 6 ÷ 3

   a. −, +, ×, ÷

   b. +, −, ×, ÷

   c. +, −, ÷, −

   d. +, ×, ÷, −

10. Which is the correct equivalent expression?

    a. 0.2, ½, 20%

    b. 0.1, 20/200, 10%

    c. 0.2, 1/5, 10%

    d. ¾, 75%, 0.6

11. Daniel calculated the area of the top surface of his workbench to be 1296 square inches. What is 1296 square inches converted to square feet?

    a. 10

b. 11

c. 9

d. 12

12. Which capacities are written in order from greatest to least?

    a. 9 milliliters, 4 liters, 8 gallons

    b. 9 liters, 5 milliliters, 1 liter

    c. 1 liter, 2 liters, 5,000,000 milliliters

    d. 10 gallons, 11 liter, 7 milliliters

13. How many gallons are there in 56.775 liters?

    a. 15

    b. 25

    c. 26

    d. 27

14. Luke attends an 80-minute sport class one day each week. Last week, he practiced the following sports classes: 20 minutes on tennis, 35 minutes on basketball, and 25 minutes on football. Which graph is best to represent the data?

    a. Line graph

    b. Circle graph

    c. Histogram

    d. Box and whisker

15. What is the mean of the following numbers? 12, 11, 14, 10, 8, 13, 11, 9

    a. 11

    b. 10

    c. 14

    d. 8

16. Which of the following measures can be found by arranging the data values in an increasing or decreasing order?

   a. Mean

   b. Mode

   c. Median

   d. None of these

17. $(-2 \times 8) - (-3 \times (-4)) =$ _____

   a. 16

   b. 4

   c. -4

   d. -28

18. $-3415 \times 10 \times 0 =$ _____

   a. 0

   b. 10

   c. -15

   d. -341500

19. 20 students were asked how many hours they spend studying. The results are 0, 0, 1, 1, 1, 2, 2, 2, 2, 2, 3, 3, 3, 4, 4, 4, 4, 7, 7, 7

How many of the students' study 4 or more hours?

   a. 4

   b. 7

   c. 2

   d. 20

20. In question 19, what is the median study hours

    a. 10

    b. 11

    c. 2.5

    d. 12

21. In question 19, what is the pick hour?

    a. 3

    b. 7

    c. 4

    d. 2

22. In question 19, what is the range?

    a. 5

    b. 6

    c. 7

    d. 8

23. The perimeter of a rectangle with lengths 18.8cm and width 7.61cm.

    a. 16.68 cm

    b. 52.82 cm

    c. 14.68 cm

    d. 28 cm

24. Write expression describing the sentence. Three times a number minus 5.

    a. 5x + 3

b. 3x - 5

c. 3x + 3

d. 8x + 3

25. Find the number if the quotient of a number and 5 is 25.

    a. 125

    b. 75

    c. 50

    d. 12.5

26. Which pair of expressions are equivalent?

    a. 2(3 + x) and 3 + 2x

    b. 3x – y and 3 (x – y)

    c. 2x + 4 and 2(2x + 2)

    d. 3x + 9y and 3(x + 3y)

27. Three times a number minus 15 is 12. What is the number?

    a. 5

    b. 6

    c. 9

    d. 8

28. The chance of some event happening is 1.3. This statement is _____.

    a. True

b. False

c. Sometimes true

d. Sometimes false

29. John reads a maximum of 50 pages an hour. Which inequality represent his speed. Use x for the speed?

a. $x < 50$

b. $x > 50$

c. $x \leq 50$

d. $x \geq 50$

30. The area of a trapezoid is given by the formula $2x^2 + 5x + 2$. What is the area when $x = 2$ meters.

a. 20 m²

b. 16 m²

c. 10 m²

d. 14 m²

31. Which number satisfy the inequality: $0.01x + 0.4 > 6$

a. 5

b. 10

c. 1, 000

d. 50

The histogram below shows the weekly income of students in dollars. Answer questions 32 – 38 questions based on the histogram.

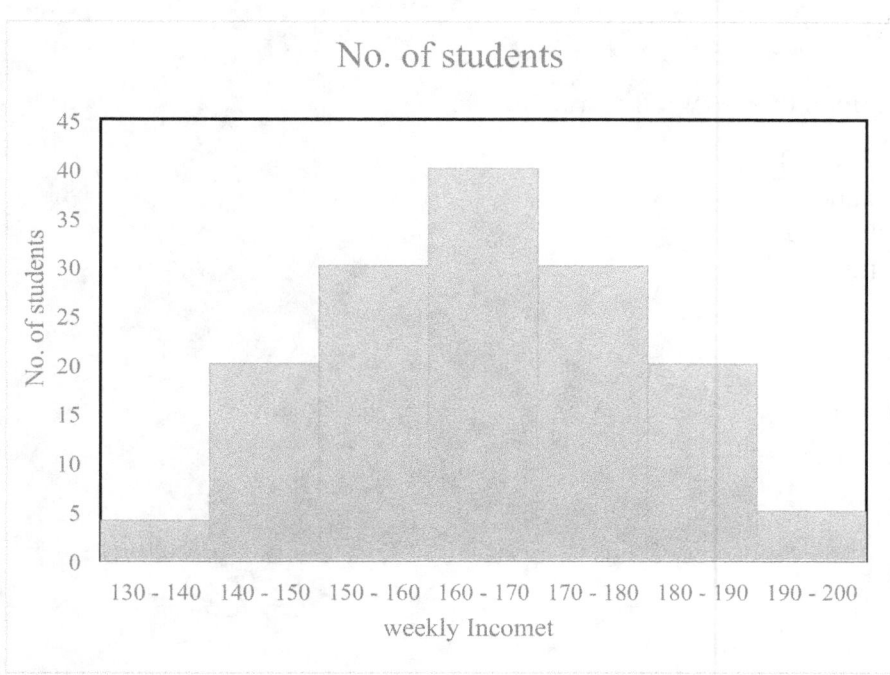

32. How many students are working? _____

33. How many students earn $170 or more? _____

34. How many students earn $160 or less? _____

35. How many earn between $140 and $200? _____

36. What is the maximum income range? _____

37. What was the most common income range? _____

38. How many were overpaid? Consider $170 or more as overpay _____

39. Which comparison is true?

   a. 40 % > 2/3
   b. 2% = 0.02
   c. 4% = 4/10
   d. 0.03 = 30%

40. The inequality 0.05x – 4y < 0 is true for:

   a. X =0 and y = 0
   b. X = 100 and y = 1
   c. X = 0 and y =2
   d. X = 0 and y = -1

ANSWERS

CHAPTERS AND SECTIONS

1.1.1.
   a. 100.7
   b. 11,051
   c. 15.936
   d. 6.522
   e. 5,035.4
   f. 3.725
   g. 530.57
   h. 39.0126
   i. 2.375267
   j. 1576.995
   k. 1,100,000
   l. 1,065,607.59

1.1.2
   a. 433.7
   b. 1,517.22
   c. 71.836
   d. 4.226
   e. 4739.1
   f. 0.529
   g. 337.85
   h. 38.483
   i. 6.604347
   j. 1116.143
   k. 899999.8
   l. 1314.89

1.2.

1.2.1 a, 2 b, 13 c, 5 d, 3 e, 5 f, 213 g, 22

1.2.2 a, 11 b,473 c, 188 d, 28 e, 24 f, 76 g, 29

1.2.3 a, 16.2 b, 6.9 c, 178.5 d, 52.9 e, 66.7 f, 79

1.3.1
   a. 10
   b. 31.5

315

c. 2.07
d. 13.32
e. 67.9
f. 59.29
g. 35
h. 201
i. 72
j. 580.8
k. 243
l. 980.1

1.3.2

1. 3.72
2. 0.096
3. 12.88
4. 19.866
5. 23.997
6. 0.2871
7. 291.06
8. 436.2441
9. 1,897.7152
10. 12.33
11. 0.123
12. 0.0321
13. 0.002645
14. 0.001
15. 0.00001
16. 0.000201
17. 0.01221
18. 0.69832
19. 9.0496
20. 10.37124

1.4.

a. 4
b. 40
c. 9
d. 83
e. 21
f. 47
g. 14
h. 58
i. 29
j. 18
k. 23
l. 34

1.5.

a. 537.5
b. 205
c. 21230
d. 987.6
e. 3.424
f. 0.0134
g. 0.064
h. 0.00031
i. 5.375
j. 0.0205
k. 0.02123
l. 0.0009876
m. 342.4
n. 134
o. 6400
p. 310
q. 4.005
r. 1
s. 1
t. 0

1.6

a. 16
b. 40
c. 30
d. 40
e. 36
f. 48
g. 72
h. 12
i. 0
j. 4
k. 8
l. 2
m. 5
n. 1
o. 4
p. 4
q. 3
r. 0

1.7.1

1. 3.1
2. 4.1
3. 3.4

4. 2.5
5. 0.5
6. 4.9
7. 6.92
8. 0.9
9. 0.56
10. 14.81
11. 2.4
12. 2.45
13. 3.8
14. 2.01
15. 400
16. 200
17. 1,000
18. 100
19. 2.5
20. 0.25

1.7.2

1. 7.2
2. 230
3. 232
4. 0.2
5. 2.3
6. 0.4
7. 840
8. 2.42
9. 21
10. 135.68
11. 17
12. 7.2
13. 2.5
14. 6.2
15. 0.24
16. 12.4
17. 2.5
18. 6.5
19. 3.25.
20. 2500

1.8.1

1. <
2. >
3. <
4. >
5. <
6. <
7. >
8. =
9. <

1.8.2

1. 0.08, 0.3, 1.2, 3.45
2. 0.04992, 0.4325, 2.0432, 4.321
3. 5.05, 50.05. 55.5, 55.505
4. 11.11, 21.01, 21.011, 21.11
5. 29.9993, 30.00009, 30.0003, 30.0016
6. 0.4132, 0.4213, 0.4312, 0.4321
7. 0.0013, 0.04213, 0.4312, 0.8

1.9.

1. 111.475 miles
2. $80.01
3. 6452.25 miles
4. 5ft.
5. 39.2ft.
6. $9.999
7. $271.25
8. $278.6
9. 30.1075lb.
10. 19.5ft.
11. 0.415lb.
12. 135.6 miles
13. 340.56gal.
14. 3.2
15. 1
16. 88.56
17. 19.92lb.
18. 49.36in.
19. 2.54m.
20. 24
21. 1163.4
22. 30

23. 8.475 miles
24. Messi
25. 40.25 miles

1.10. CHAP. 1 Test
1. c
2. a
3. b
4. b
5. d
6. a
7. d
8. c
9. c
10. a
11. a
12. a
13. b
14. a
15. a
16. c
17. a
18. d
19. c
20. b

2.1.1
1. 1
2. 11/5
3. 2
4. 26/11
5. 0
6. 6
7. 8/5
8. 21/5
9. 0
10. 2/5

2.1.2
1. 29/15
2. 19/35
3. 23/20
4. 13/5
5. 17/2
6. 55/24
7. 40/7
8. 16/35
9. 34/15
10. 2
11. 59/30
12. 8/25
13. 9
14. 179/30
15. 169/30
16. 109/120
17. 547/140
18. 1467/114
19. 3
20. 323/140

2.2.
1. 0
2. 1/2
3. 1/2
4. 11/2
5. 13
6. 1/2
7. 1
8. 0
9. 12
10. 12

2.3.1
a. Proper
b. Proper
c. Improper
d. Proper
e. Improper
f. Improper
g. Improper
h. Proper
i. Proper
j. Proper

2.4
1. $\dfrac{103}{20}$
2. $\dfrac{101}{5}$
3. $\dfrac{127}{25}$
4. $\dfrac{147}{25}$

5. $\dfrac{17}{10}$
6. $\dfrac{202}{100}$
7. $\dfrac{41}{5}$
8. $\dfrac{5}{4}$
9. $\dfrac{23}{4}$
10. $\dfrac{82}{11}$
11. $\dfrac{103}{33}$
12. $\dfrac{1291}{213}$
13. $\dfrac{45}{21}$
14. $\dfrac{78}{11}$
15. $\dfrac{175}{8}$
16. $\dfrac{187}{33}$
17. $\dfrac{177}{22}$
18. $\dfrac{8255}{110}$
19. $\dfrac{37}{9}$
20. $\dfrac{937}{3}$

2.5.

1. $1\dfrac{1}{4}$
2. $2\dfrac{1}{20}$
3. $3\dfrac{2}{3}$
4. $1\dfrac{4}{21}$
5. $2\dfrac{3}{22}$
6. $1\dfrac{6}{27}$
7. $8\dfrac{7}{9}$
8. $1\dfrac{11}{47}$
9. $1\dfrac{1}{34}$
10. $13\dfrac{9}{15}$
11. $17\dfrac{11}{17}$
12. $22\dfrac{10}{27}$
13. $15\dfrac{9}{25}$
14. $5\dfrac{23}{200}$
15. $21\dfrac{5}{20}$
16. $42\dfrac{8}{21}$

2.6.1.

1. $10\dfrac{1}{6}$
2. $11\dfrac{6}{13}$
3. $2\dfrac{2}{7}$
4. $4\dfrac{1}{6}$
5. $4\dfrac{3}{6}$
6. $13\dfrac{1}{12}$
7. $4\dfrac{14}{15}$
8. $9\dfrac{2}{11}$
9. $5\dfrac{5}{7}$
10. 26

2.6.2

11. 22
12. $3\dfrac{5}{12}$
13. $4\dfrac{17}{42}$
14. $4\dfrac{7}{30}$
15. $4\dfrac{5}{6}$
16. $22\dfrac{31}{60}$
17. $6\dfrac{1}{5}$

18. $11\frac{5}{22}$

19. $19\frac{16}{21}$

20. $4\frac{19}{42}$

2.7.1

1. 7
2. 5
3. 5/4
4. 4
5. 28
6. 105
7. 264
8. 605/6
9. 180
10. 89,673/27
11. 50
12. 0
13. 1332/5
14. 356,775/29
15. 40,000

2.7.2

1. 2/15
2. 9/20
3. 33/2
4. 38/192
5. 6/49
6. 2/35
7. 13/15
8. 32/49
9. 133/150
10. 169/125
11. 931/3750
12. 76/18
13. 7
14. 7/12
15. 17/14
16. 16100/361
17. 0

2.8

1. 52/3
2. 459/10
3. 13439/120
4. 2622/25
5. 321/22
6. 46/3
7. 51/4
8. 968/9
9. 323/12
10. 33/4

2.9.1

1. 84
2. 162
3. 4400
4. 38
5. 0.83
6. 2
7. 36,960
8. 42,240
9. 0.5
10. 126,720
11. 6
12. 2
13. 0.00
14. 0.01
15. 6
16. 1,320
17. 100
18. 0.87
19. 0.32
20. 2

2.9.2

1. 15
2. 18
3. 144
4. 7
5. 3.5
6. 3
7. 8
8. 64
9. 2
10. 28
11. 3
12. 12

13. 24
14. 128
15. 40
16. 7

2.10.1

1. 20
2. 78
3. 8/3
4. 36
5. 72/7
6. 16
7. 66
8. 30
9. 72
10. 351/42
11. 21
12. 0
13. 55
14. 58
15. 20

2.10.2

1. 0
2. 2/9
3. 11/6
4. 57/2
5. 1
6. 40/7
7. 108
8. 16
9. ¾
10. 1
11. 21/38
12. 8/19
13. 20/9

2.11

1. 10/11
2. 176/27
3. 329/144
4. 342/215
5. 56/27
6. 306/445
7. ½
8. 214/33
9. 133/138
10. 217/114

2.12.1.

1. 2,3,5,7, 11, 13, 17, 19, 23, 29

2. 4,6,8,9,14,15,16,18,20,21,22,24,25, 26,27, 28, 30, 32, 34

3. a. $2 \times 2 \times 3$

b. 3×3

c. $2 \times 3 \times 3$

d. $2 \times 2 \times 2 \times 2$

e. 2×17

f. $3 \times 5 \times 5$

2.12.2.

1. 1,2,4,8
2. 1, 11
3. 1,2,3,4,6,12
4. 1,2,4,7,14, 28
5. 1,2,4,13, 26,52
6. 1,2,3,6,17, 34,51, 102
7. 1,2,4,17,34,68
8. 1,2,3,5,6,7, 10,14,15,21,30,35,105, 210
9. 1,17
10. 1, 59
11. 1, 43
12. 1, 3, 17, 51
13. 1, 3, 43, 129
13. 1,3,43,129
14. 1,3,5,15,25, 75
15. 1,2,3,6,25,50,75,150
16. 1,2,3,4,5,6,8,10,12,15,20, 24,30,40,60,120

2.13.

1. 1

2. 6
3. 1
4. 10
5. 18
6. 1
7. 9
8. 2
9. 100
10. 12
11. 24
12. 16
13. 42
14. 4
15. 1
16. 9

2.14.
1. 110
2. 120
3. 80
4. 48
5. 180
6. 60
7. 48
8. 120
9. 120
10. 720
11. 432
12. 1,224

13. 60
14. 84
15. 60
16. 30

2.15.
1. 1/9
2. 1/6
3. 14/17
4. ½
5. 17
6. 1/7
7. 1/3
8. 1/9
9. 10/19
10. 4
11. 5
12. 4/7
13. 6/7
14. 2
15. 7/16
16. 5/12
17. 5/4
18. 1/6
19. 1/5
20. 3/8

2.16
1. 2/25
2. 11/4

3. 7
4. 56/3
5. 1/5
6. 10
7. 1/9
8. 243/28
9. 1/5
10. 3/5
11. 2/5
12. 1/180
13. 20/9
14. ¾
15. 2/3
16. 1/36
17. 9/2
18. 27
19. 2/3
20. ½

2.17
1. 1,033/798
2. 9/11
3. 11317/720
4. 19/15
5. 347/120

2.18.1
1. >
2. >
3. >

4. =
5. =
6. <
7. <
8. <
9. <
10. >
11. <
12. >
13. =
14. >
15. >
16. <
17. >
18. =
19. <
20. >
21. >
22. >

2.18.2

1. $\frac{1}{10}, \frac{2}{5}, \frac{1}{2}$
2. $\frac{1}{2}, \frac{6}{10}, \frac{2}{3}$
3. $\frac{1}{10}, \frac{2}{5}, \frac{1}{2}, \frac{7}{4}$
4. $\frac{1}{2}, \frac{4}{5}, \frac{5}{6}, \frac{6}{7}$
5. $\frac{1}{10}, \frac{2}{5}, \frac{7}{11}, \frac{13}{12}, \frac{7}{4}$
6. $\frac{14}{5}, \frac{25}{6}, \frac{10}{2}, \frac{30}{4}, \frac{64}{7}$
7. $1\frac{4}{5}, 2\frac{5}{6}, 3\frac{3}{4}$
8. $\frac{1}{2}, 6\frac{5}{6}, 9\frac{7}{5}, 10\frac{1}{4}, 10\frac{2}{5}$
9. $1\frac{11}{5}, 2\frac{2}{5}, \frac{15}{16}, 3\frac{6}{5}, 7\frac{1}{2}$
10. $\frac{18}{7}, \frac{200}{56}, \frac{215}{36}, 23\frac{6}{5}$
11. $\frac{6}{5}, 2\frac{2}{5}, \frac{15}{16}, 1\frac{11}{5}, 10$
12. $\frac{101}{51}, 2\frac{2}{51}, 1\frac{17}{16}, 2\frac{6}{5}, 5$

2.19

1. 20
2. 60
3. June 14
4. 8
5. 2
6. 1/12
7. 3/7
8. 1/10
9. 18
10. 3
11. 6
12. 12
13. 4,000
14. 55/6
15. 8
16. 1/3
17. 161/8
18. 10/9
19. 181/6
20. 7/15
21. 97.5
22. 130
23. 24
24. 26
25. 5,395
26. 4/3 ft.
27. No
28. Yes
29. 14
30. 144
31. 8
32. 3/8

33. 3/32
34. 48
35. 200
36. 30
37. Yes
38. 40
39. 1/15
40. 11
41. 1/30
42. 60
43. 36
44. 60

3.1.1.

1. 29, 14
2. 9: 6
3. 17: 14
4. 35
5. pencils: pens (7/9)
6. Male teachers to all teachers 14:26
7. Residents living in a house to all residents: 8/10
8. Human nose to human eyes: 1/2
9. Windows to doors: 16/4
10. Chairs: tables 64/8
11. Baseball to basketball fields: 3/7

45. 1/15

2.20 test 2

1. d
2. b
3. b
4. a
5. c
6. C
7. C
8. C
9. A
10. B
12. Car to airplane travelers: 750/250 (answer varies)
13. Language art to science: 40/35. Science to math: 35/25. Math to language art: 25/40 (answer varies)
14. Younger than 25 to all age: 7/10
15. 5/7
16. 2/3, 2/5, & 3/5

3.1.2.

1
a. 3:4
b. 1:4
c. 1:2
d. 1:5
e. 4:3

11. A
12. A
13. B
14. D
15. C
16. A
17. D
18. A
19. C
20. C

    f. 1:10
    g. 3:1
    h. 5:2

2
a. ¼
b. 1/5
c. 2/3
d. 1/50
e. 4
f. 1/3
g. 1/3
h. 3/100
i. ¼
j. 10/1
k. 1/8

3.2

1. 2:1

2. 24miles/3hours
3. 8miles/1hour
4. 5miles/hour
5. 30miles/1gal.
6. 5ft./1second
7. 40 miles
8. 20m/s
9. 9 pages
10. Sally

3.3.1 (Answer varies)

1. 4/10, 6/15
2. 6/8, 9/12
3. 10/12, 15/18
4. 22/14, 33/21
5. 6/24, 9/36
6. 4/38, 6/57
7. 2/26, 3/39
8. 10/34, 15/51
9. 6/14, 9/21
10. 25/117, 100/468
11. 62/140, 93/210
12. 58/134, 87/201

3.3.2

1. 18
2. 5
3. 49
4. 473
5. 108
6. 25
7. 2
8. 4
9. 9
10. 10
11. 8
12. 9
13. 336
14. 7
15. 34
16. 1
17. 36
18. 144
19. 1
20. 270
21. 18
22. 10
23. 22
24. 2
25. 3
26. 13
27. 6
28. 135
29. 4
30. 4

3.4.1

1. 15 minutes
2. 12
3. 6
4. 20
5. 4

3.4.2

1. Mark

1	2	3
8	16	24

Bob:

1	2	3
8.5	17	25.5

2. $68

3. $76.5

4. 12 hours

5. No

6. Mark: (1, 8), (2, 16), (3, 24)

Bob: (1, 8.5), (2, 17), (3, 25.5)

3.5.1

1. 1/4
2. ½
3. 3/25
4. 17/25
5. ¾
6. 43/50
7. 1/100
8. 1
9. 1/8
10. 21/500

3.5.2

1. 5%
2. 30%
3. 50%
4. 25%
5. 25%
6. 75%
7. 10%

8. 25%
9. 66.7%
10. 40%
11. 90%
12. 110%
13. 62.5%
14. 8%

3.6.1
1. 0.43
2. 0.23
3. 0.221
4. 1
5. 0.0201
6. 0.75
7. 5.41
8. 0.0001
9. 1.01
10. 0.01
11. 0.065
12. 0.999
13. 0.101
14. 0.0028

3.6.2
1. 85%
2. 1%
3. 10%
4. 123%
5. 6500%
6. 467%
7. 2123%
8. 99%
9. 990%
10. 9990%
11. 2.1%
12. 208%

3.7.1
1. 0.75, 75%
2. 0.375, 37.5%
3. 0.25, 25%
4. 0.125, 12.5%
5. 0.1, 10%
6. 0.125, 12.5%
7. 0.111, 11.1%
8. 0.1, 10%
9. 0.01, 1%
10. 0.1. 10%

3.7.2
1. 25%, 0.25
2. 35%, 0.35
3. 83.3%, 0.833
4. 2%, 0.02
5. 125%, 1.25
6. 50%, 0.5
7. 5%, 0.05
8. 12.5%, 0.125
9. 33.3%, 0.333
10. 40%, 0.4

3.8
1. 53
2. 36
3. 5%
4. 12.5%
5. 40.5
6. 232
7. 226.8
8. 243
9. 500
10. 330

3.9
1. =
2. <
3. <
4. >
5. <
6. <
7. =
8. =
9. <
10. <
11. 1/8, 1/7, 0.2. 0.56
12. 3%, 1/25, 0.07
13. 25%, 3/9, 0.4
14. 89%, 110%, 1.25
15. 1/3, 2/3, ¾
16. 37.5%, 3/7, 0.5, 3/5
17. 9%, 0.9, 100%, 11/10
18. 4/12, 26%, 0.31, 0.6
19. 0.6, 10/15, ¾, 6/7

326

20. The same numbers

3.10

1. 300
2. 16
3. 40
4. 8
5. 42
6. 60
7. 10
8. 20
9. 48
10. 14
11. 210
12. 120
13. 75
14. 60
15. 60
16. 810
17. 290
18. 180
19. 140
20. 0

3.11.

1. $1058
2. $33.32
3. $105,000
4. 240
5. No
6. 3
7. 83.3%
8. 0.9 liters
9. $144
10. $190
11. 190 days
12. $ 7.50
13. 50%
14. 216
15. $4800
16. 9.6 gallons
17. 75%
18. 40%
19. $60
20. $240

3.12.

1. a. 2/3, b. 7/4, c. 9/13, d. 13/9, e. 5/6, f. 6/11, g. 5/11
2. the 64 –ounces container
3. a. 10, b. 18, c. 2, d. 9, e. 5, f. 14, g. 15
4. 54 minutes
5. 135 pages
6. 20
7. 12
8. $7.50
9. $20
10. $182.25
11. 10 meters
12. $445.6
13. 95,000

3.13 (Test)

1. c
2. d
3. b
4. b
5. c
6. b
7. a
8. d
9. b
10. b
11. c
12. b
13. d
14. b
15. a
16. a
17. a
18. d
19. b
20. b
21. b
22. c
23. c
24. b
25. b

4.1

1. 3^3
2. 5^5
3. $(0.4)^4$
4. $(1/2)^3$
5. 512
6. 32

7. 0.027
8. 1/8
9. 4 × 4
10. 0.1 × 0.1 × 0.1
11. 1 × 1 × 1 × 1 × 1
12. (1/4) × (1/4) × (1/4)
13. 1
14. 1
15. 32
16. 1,000
17. 0
18. 1/8
19. 5.29
20. 1.0201

4.2

1. 6
2. 21
3. 76
4. 22
5. 58
6. 2.5
7. 10
8. 1
9. 4
10. 95
11. 96
12. 74
13. 90
14. 12
15. 275
16. 848
17. 182
18. 121
19. 248
20. 9
21. 45

4.3.1

1. x
2. a, x
3. no
4. a, b, c
5. x, y
6. x, y, p

4.3.2

1. Equation
2. Expression
3. Expression
4. Expression
5. Equation

4.3.3

1. ×, -
2. ×, +, -
3. ×, ÷
4. ×, +
5. ×, ÷
6. ×, -
7. ×, ÷, -
8. ×, +
9. ×, -, ÷
10. ×, +, ÷
11. ×, ÷, -
12. ×, +, -

4.4

1. 19
2. 6
3. 7
4. 2
5. 1
6. -1
7. 5
8. -310
9. 5025
10. -99
11. -4
12. 12
13. 5
14. -25
15. 4
16. 13
17. -10
18. -308
19. 43/25
20. 1
21. -2/3
22. 1
23. 25/3
24. 15
25. 3.5

26. 19/7
27. -19/4
28. 18
29. -1
30. 8

4.5

1. 7x + 8
2. X -5
3. (x/4) – 3 = 8
4. 4x
5. 2x + 6
6. x + 5
7. 2x = 10
8. x – y = 5
9. 5 – x = 20
10. X + 3 = 8
11. 2x – 10 = 5
12. 2x = 20
13. 2x + 1 = 25
14. 3 × 14 = 42
15. (1/6) x = 5

4.6.

1. Commutative property of multiplication
2. Commutative property of addition
3. Commutative property of addition
4. Commutative property of multiplication
5. Associative property of addition
6. Associative property of addition
7. Distributive property of x over +
8. Distributive property of x over +
9. Distributive property of x over +
10. Associative property of multiplication
11. Commutative property of multiplication
12. Associative property of multiplication

4.7.

1. Y +12
2. 24y
3. 7 – x
4. 2x + 2y
5. 3y
6. 0
7. 29 – x
8. 6 – x
9. 44 or 4×11
10. 2x – 4
11. 60 or 3×20
12. 5x + 5y
13. Yes, identity property of addition
14. Not equivalent
15. Yes, associative property of addition
16. Not equivalent
17. Yes, associative property of multiplication
18. Not equivalent
19. Yes, identity property of multiplication
20. Not equivalent

4.8.1

1. 5×2 +5×4
2. 3x +6
3. 2x + 8
4. 30 + 24y
5. 12x + 20
6. 16x + 24y
7. 48x + 6y

8. $4x + 5y$
9. $8 + 4x$
10. $x + 2y$
11. $4x + 6y + 8w$
12. $9x + 9y + 12z$
13. $x + 2y + 3z$
14. $6 + 10x + 4y$
15. $6x + 9 + 15y$

4.8.2.

1. $3(5 + 7)$
2. $7(1 + 7)$
3. $8(2 + 3)$
4. $4(x + 4)$
5. $5(x + 2)$
6. $12(5 + 2x)$
7. $7(8x + 7)$
8. $3(2x + 3y)$
9. $5(14x + 5y)$
10. $4(2x + 9y)$
11. $2(2x + 3y + 7z)$
12. $3(x + 2y)$
13. $2(10x + 9y + 18)$
14. $2(x + 2y + 4)$
15. $5(x + 5y + 15)$

4.9.1.

1. 20
2. 72
3. 129
4. 24
5. 42
6. 39
7. 56
8. 36
9. 20
10. 63

4.9.2.

1. =
2. >
3. <
4. =
5. =
6. =
7. >
8. <
9. >
10. =

4.9.3

1. $20x$
2. $4x + 8$
3. $9x$
4. $16x + 13$
5. $3x + 20$
6. $15x + 8$
7. $17x + 6$
8. $3x + 8$
9. $7x$
10. $19x + 41$

4.9.4.

1. $3x + 5y$
2. $20x + 2y$
3. $17x + 8$
4. $26y + x$
5. $18x + 4y$
6. $11x + 2y$
7. $6x + 27y$
8. $6x + 32y$
9. $14x \; 12y + 8$
10. $18x + 8y$

4.10.1.

1. 7

2. 16
3. 16
4. 30
5. 40
6. 50
7. 100
8. 100
9. 100
10. 200

4.10.2.

1. 2
2. 6
3. 17
4. 49
5. 10
6. 12
7. 50
8. 10
9. 60
10. 12

4.11.

1. 4
2. 4
3. 16
4. 11
5. 12
6. 3
7. 10

8. 7
9. 9
10. 10
11. -5
12. 3
13. -4
14. 3
15. -8
16. -9
17. – 12
18. 13
19. - 8
20. 10

4.12.1.
1. False
2. True
3. False
4. True
5. False
6. False
7. True
8. False
9. True
10. True
11. False
12. True
13. True
14. False
15. True
16. False
17. True
18. True
19. True
20. False

4.12.2.
1. 3
2. 2
3. -1
4. 4
5. -5
6. -5
7. -10
8. 16
9. -14
10. 1
11. -2
12. 0
13. 4
14. – 5
15. - 6
16. 0
17. 2
18. 1
19. 4
20. 2

4.13
1. 20
2. 10
3. 120
4. 25
5. -5
6. 6
7. 18.84
8. 6

4.14
1. 12
2. 27
3. 11
4. 10
5. 12
6. -4
7. 5
8. 14
9. 11
10. 6, 7
11. 408
12. 6

4.15 (Test)
1. c
2. d
3. b
4. b
5. c
6. b

7. a
8. b
9. c
10. d
11. c
12. c
13. d
14. c
15. b
16. d
17. c
18. c
19. b
20. d
21. a

5.1.

1. False
2. True
3. True
4. False
5. False
6. True
7. False
8. True
9. True
10. True

5.2.1.

1. 6, 7, 8, 9
2. 16, 24, 32, 40
3. 28, 53, 78, 103, 128
4. 3, 1, -3, -5, -7
5. 6
6. -4
7. 19
8. 6
9. 17
10. 44

5.2.2

1. 5x
2. X + 3
3. 2x + 1
4. 3x + 2
5. 4x - 3

5.3.1

1. Arithmetic
2. Neither
3. Arithmetic
4. Arithmetic
5. Geometric
6. Geometric
7. Arithmetic
8. Neither
9. Arithmetic
10. Geometric
11. 37
12. 243
13. 4
14. 3
15. 7
16. 1/2
17. 625
18. 77
19. 5.25
20. 51
21. 25
22. 32
23. 21
24. 11.8
25. 1
26. 22.8
27. 0.4
28. 6
29. 6
30. 343

5.4.

1. No
2. Yes
3. Yes
4. Yes
5. No
6. No
7. Yes
8. No
9. No
10. Yes
11. 9, 10
12. 0, 1
13. 10, 11
14. 9, 10
15. 0, 1

16. 4, 5

17. 10, 11

18. -1, -2

19. 3, 4

20. 0, 1

5.5.

1. $X \geq 5$
2. $X \leq 200$
3. $X \leq 45$
4. $X \geq 6$
5. $X < 2$
6. $X > 10$
7. $X \leq 3$
8. $X > 6$
9. $X \leq 4$
10. $2 < x \leq 7$

5.6.1.

1. $X \geq 3$
2. $X \leq 2$
3. $X > 5$
4. $X < 4$
5. $X \geq 0$
6. $X \leq -2$
7. $X > -3$
8. $X - 1 \leq 3$
9. $X + 2 \geq 1$
10. $X - 2 < 1$

5.6.2.

1. $X < 0$
2. $X \geq -1$
3. $X > -2$
4. $X \leq 3$

5.7.

1. $X < 6$
2. $X > 2$
3. $X \leq 2$

4. $X \geq 2$
5. $X \leq -14$
6. $X < 2$
7. $X < 11$
8. $X < 9$
9. $X > -7$
10. $X \leq -4$
11. $X > 14$
12. $X > -3$
13. $X > -15$
14. $X \leq -5$
15. $X > 5$
16. $X < 1$
17. $X < 6$
18. $X < 2$
19. $X < 4$
20. $X \leq -4/3$

5.8.

1. $2x + 5 < 11, x < 3$
2. $5x \leq 30, x \leq 6$
3. $X \geq 75$
4. $15x \geq 1,500, x \geq 100$
5. $3x + 3 > 102, x = 34$
6. No
7. $3.5x < 311.5, 89$
8. No
9. The difference of a number and three is less than 7
10. $X + 6 > 160$

5.9 (Test)

1. b
2. c
3. c
4. d

5. b
6. a
7. c
8. c
9. c
10. d
11. a
12. d
13. b
14. a
15. a
16. c
17. a
18. b
19. a
20. d

6.1.

1. False
2. True.
3. False
4. False
5. True
6. False
7. False
8. False
9. True
10. True
11. True
12. False
13. True
14. True
15. False
16. False
17. False

18. False
19. True
20. True
21. True
22. False

6.2

1. 20m^2
2. 35 ft^2
3. 6 ft^2
4. 30 ft.
5. 24m.
6. 10yd^2
7. 30m^2
8. 169 in^2
9. 4ft.
10. 9yd.
11. The parallelogram because it has short height
12. 45 ft^2
13. 20 ft^2
14. 80 in^2
15. 112 m^2

6.3.

1. 7m^2
2. 7m
3. The area is quadrupled
4. (1/16) times the original area
5. 6m^2
6. The perimeter doubles
7. The area quadruples
8. 6.96 ft^2
9. 8in.
10. 14ft. or 16ft.

6.4.

1. 104 cm^2
2. 30in
3. 10m
4. 2ft, 4ft
5. 20yd^2
6. 3m^2
7. 172.8in.
8. 5184 in^2
9. The area doubles
10. The area is 9 times bigger.

6.5.

1. d = 2r
2. 10in.
3. 1m.
4. 10cm.
5. 31.4ft.
6. 12.56 units
7. Yes
8. 3.14, yes
9. 200.96yd^2
10. 254.34ft^2
11. 5ft.
12. 10m.
13. 100πin^2
14. 20π units
15. 1 unit

6.6.

1. 42 m^2
2. 432 m^2
3. 88.26 ft^2
4. 27 in.^{2}
5. 20.375cm^2
6. 28m^2

6.7.

1. 5.5ft.
2. 150m
3. 37m
4. 400in.
5. 176m^2
6. 30ft.
7. 49m^2
8. $588
9. 4ft.
10. $4
11. 2,000m.
12. 20in.
13. 25m.

6.8.

1. b
2. c
3. a
4. d
5. d
6. b
7. c
8. b
9. a
10. c
11. c
12. b
13. a
14. b
15. a
16. c
17. c
18. a
19. d
20. c

7.1

1. 54yd^2

2. 148 m^2
3. 130 in^2
4. 500ft.
5. 112 m^2
6. 20π ft^2
7. 28π ft^2
8. 78 m^2
9. 83 m^2
10. 7.8 gallons
11. 92 m^2
12. 24 in^2
13. 96 m^3
14. 132π m^2

7.2.

1. A square pyramid
2. A triangular prism
3. A cube
4. A triangular pyramid
5. Pentagonal prism
6. A cylinder

7.3.

1. 200 m^3
2. 120 in^3
3. 96 m^3
4. 180π m^3
5. 512 ft^2

8.1.

1. Yes
2. Yes
3. Yes
4. No
5. No
6. Yes
7. No
8. No
9. Yes

6. 56 in^3
7. 20 cm^3
8. 48 yd^3
9. 225.48 in^3
10. The volume be 8 times more

7.4.

1. 72mm^3
2. 125in^3
3. 3ft.
4. 6m.
5. 3in.
6. 64yd^3
7. 56m^3
8. 628m^3
9. 4yd.
10. 267.9cm^3

7.5.

1. 1272 ft^2
2. $1273
3. 220 in^3
4. 10cm
10. No
11. Yes
12. No
13. Yes
14. Yes
15. Yes
16. No
17. Yes
18. No
19. No
20. Yes

8.2

5. 166 in^2
6. 100
7. 720
8. 20 min.
9. 12.9 L
10. 43 gal.

7.6.

1. A
2. B
3. B
4. B
5. A
6. C
7. C
8. D
9. D
10. A
11. D
12. B
13. B
14. C
15. D
16. 10hrs.
17. 250ft^2
18. 400m^3
19. 216in^2
20. 6

8.2.1

1. 78.89
2. a. 42.57, b. 50, c. 5.93, d. 55, e. 42, f. 6.96
3. a. 3, b. 7
4. 17
5. 16
6. 5

7. $225
8. $4
9. No
10. 15

8.2.2

1. a. 39, b. 50, c. 4, d. 60.5
2. the 7th number
3. The 5th number plus the 6th number divided by 2
4. 23
5. 5
6. 15

8.2.3

1. Unimodal
2. Bimodal
3. No mode
4. Bimodal
5. 23
6. 13 and 20
7. 52
8. 2 and 3

8.2.4.

a. 2,4,4,6
b. 3,5,5,7
c. 3,5,5,7
d. 3,3,3

8.3.1

1. 82.5
2. 0
3. 68
4. 486
5. 15
6. 2
7. 9
8. 1, 2, 6 (answer varies)
9. 2, 2, 2, 2 (answer varies)
10. 1, 1.2, 1.4, 2 (answer varies)

8.3.2.

1. 2, 4, 6
2. 3.5, 5, 10.5
3. 3, 5.5, 8
4. 2, 3, 5
5. 5, 6, 9
6. 32, 46, 85
7. 3, 6, 7
8. 3.5, 6, 9
9. 1.5, 4, 6.5
10. 45, 63, 74
11. 37.5, 49.5, 71.5
12. 48, 63.5, 77
13. 26, 55.5, 82
14. 24.5, 27, 49
15. 12.5, 17, 33
16. 4, 4, 4

8.3.3.

1. 4
2. 7
3. 5
4. 3
5. 4
6. 53
7. 4
8. 5.5
9. 5
10. 29
11. 34
12. 29
13. 56
14. 24.5
15. 20.5
16. 0

8.3.4.

1. 1.7
2. 3.1
3. 2.2
4. 6
5. 4
6. 4.4
7. 6
8. 2.7
9. 2.7
10. 14.7

8.4.

1. 1, 19, No
2. 4, 16, No
3. 3, 21, No
4. 3, 25, No
5. 5, 13, No
6. 6, 14, No
7. 35, 85, No
8. 23, 95, No
9. 22, 92, No
10. 23, 87, No
11. 22, 81, No
12. 13, 180, Yes
13. 6, 6, No

8.5.

1. 42
2. 99
3. 62

4. 97
5. 35
6. No
7. 25%
8. 50%
9. 75%
10. 75%

8.6.

1. 3
2. 95
3. 50
4. 7
5. 10
6. 0
7. 18
8. 15
9. 23
10. 7
11. 85
12. 43.5%
13. 87%
14. 0%
15. 100%

8.7.1

1. 500
2. America
3. France
4. France
5. Ethiopia
6. 1,500
7. 500
8. 2,640
9. 360
10. 540

8.7.2

1. 200
2. 200
3. 400
4. C
5. 11.1%
6. 450
7. 50
8. 400
9. 88.9%
10. No

8.8.

1. Week 1
2. Increases
3. Week 6
4. 10 inches
5. 10 inches
6. 8.5 inches
7. Week 6
8. Week 2
9. Week 3
10. Yes

8.9.1

1. 50%
2. 90
3. 98
4. 12
5. 80º F - 99º F
6. 0º F - 19º F
7. 119º F
8. 80º F - 99º F
9. 42
10. 0%
11.

class	frequency
0 - 9	12
10-20	8
20-29	9
30-39	6
40-49	5

8.9.2

1. 150
2. 55
3. 55
4. 145
5. 190 – 200
6. 160 – 170
7. 55
8. Normal
9. 63%
10. 80%

8.10.

1. 80
2. 60
3. 40
4. Basketball
5. 60
6. 120
7. 220
8. 280

9. True
10. True

8.11.

1. 35
2. 75
3. 45
4. 26.1%
5. 10.9%
6. 43.5%
7. 230
8. 25
9. 55
10. 75
11. 23.9%
12. 80.4%
13. 52.2%
14. No
15. False

8.12.

1a. 2% b, 4/5 c, 40%

d, 0.89 e, 66%

8.13.1

1. Tossing
2. Tossing
3. Rolling
4. Spinning
5. Pulling
6. Assigning numbers
7. Coloring

8.13.2

1. On, Off
2. {head head, head tail, tail head, tail tail}
3. {1, 2, 3, 4, 5, 6}
4. {1,2,3,4,5,6,7,8}
5. {R, R, B, G, G, G, Y, Y}
6. {2, 3, 4, 5, 6, 7}
7. {Red, Yellow, Blue, Green, White, Black}

8.14.

a. 25%
b. 15%
c. 60%
d. 50%
e. 35%
f. 75%
g. 85%
h. 75%
i. 90%

8.15.

1a. 1/6 b, 1/6 c, 1/3

d, ½ e, ½ f, 1/3

2a. 2/9 b, 2/9 c, 1/9

d, 1/9 e, 1/3 f, 4/9

g, 4/9 h, 7/9 I, 7/9

8.16.

1
   a. 6
   b. 28
   c. 22
   d. Yes
   e. 11
   f. 198
   g. Median
   h. Interquartile range
   i. 33.6

2.
   a. 20.5
   b. 23.5
   c. 3
   d. No
   e. 8
   f. 22
   g. Mean
   h. Mean absolute deviation
   i. 22
   j. No

3.
   a. 3
   b. 5.5
   c. 7
   d. 5.8
   e. 5.625
   f. 40%
   g. 8
   h. 0%
   i. 100%
   j. No

4. ½

5. ½
6. ½
7. ¾
8. 5/8
9. 1/6
10. 1/3
11. 32
12. 0.08
13. 1/6

8.17. Test

1. a
2. c
3. a
4. d
5. b
6. C
7. C
8. B
9. A
10. D
11. B
12. 97.2
13. 27
14. 80
15. B
16. D
17. D
18. D
19. 12
20. D
21. B
22. A
23. B
24. 19
25. 19
26. b
27. c
28. b
29. a
30. d
31. b
32. a
33. a
34. c
35. d
36. ¼
37. 3/8
38. 7/8
39. 0
40. 1

9.1.

1. 0.1
2. 22.2
3. 1/2
4. 2.5
5. 9.8
6. 3/4
7. 1.25
8. π
9. 1/3
10. 3/8

9.2.

1. -2
2. -5
3. 3
4. -16
5. 12
6. 33
7. 100
8. -45
9. 602
10. -9,100

9.3.1

1. >
2. <
3. >
4. >
5. =
6. >
7. >
8. >
9. >
10. <

339

11. <
12. >
13. >
14. >
15. <
16. >
17. <
18. >
19. >
20. <

9.3.2
1. -2, 0, 2, 4, 6
2. -9, -4, 5, 6, 43
3. -23, -11, 0, 21, 4
4. -44, -36, -5, 3, 5, 37
5. -12, -7, -1, 2, 17
6. -33, -16, 17, 32
7. -1000, -100, -10, 10, 100
8. -999, -99, -19, -9, 0, 99
9. -3, -2, -1, 0, 1, 2, 3
10. -66, -44, -33, -32, 24, 32, 68
11. -43, -33, -32, -5, 0, 42
12. -15, -9, -8, -5, 5, 48, 80

9.4.1

1. -5
2. 0
3. -4
4. 6
5. 56
6. 15
7. -27
8. -13
9. 160
10. 0
11. -66
12. 9
13. -99
14. 78
15. -10
16. 26
17. -14
18. -100
19. -48
20. -15

9.4.2
1. -5
2. 6
3. 2
4. -4
5. 1
6. -13
7. -6

8. 0
9. 6
10. -1
11. -49
12. -8
13. -20
14. -66
15. -49
16. -11
17. -52
18. -14
19. 0
20. -122
21. -10
22. 118
23. 31
24. 148

9.4.3
1. 1
2. 5
3. 9
4. -9
5. 9
6. 39
7. 8
8. 0
9. -30
10. 4

340

11. -77
12. 8
13. -9
14. 1
15. 12
16. -30
17. 120
18. 0

9.4.4.
1. -1
2. 1
3. -3

9.6
1. 4
2. 3
3. 2
4. 1
5. 4
6. Y-axis
7. X-axis
8. Origin
9. 2
10. 4

9.7.1

4. -45
5. 4
6. -5
7. -2
8. -1
9. 0
10. 2
11. -5
12. 1
13. 1
14. -1
15. 1
16. 2

9.5.
1. A (1, 2)
2. B (3, 5)
3. C (3, 6)
4. D (2, -6)
5. E (4, -5)
6. F (-1, -3)
7. G (-4, 2)
8. H (-3, 5)
9. I (-4, 6)
10. J (2, 0)

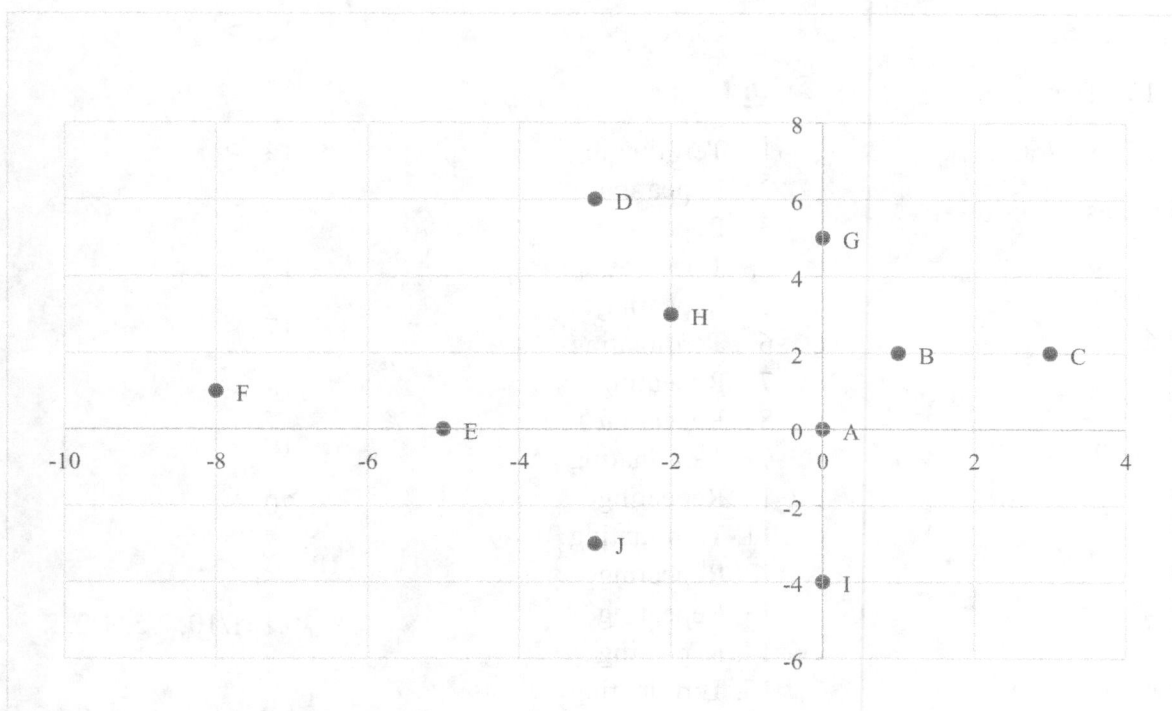

9.7.2.
1. (4, -2)
2. (-3, 5)
3. (4, 4)
4. (0, 2)

9.8.
1. Y =10
2. Y= 12, 21
3. X = 2, y = 1, 2, 3
4. x = 0, 2, y = 1, 2
5. y = -12, -8
6. x = 2, y = 1, 2, 3
7. x = 2
8. y = 0, -2, -6

9.9.

5. (4, 4)
6. (-5, 1)
7. (-8, 4)
8. (4, -8)
9. (5, 0)

1. a, (4,2) b. (-1, 2)
   c, (-1, 1) d, (-7,0)
2. a, (4, 3) b, (2, 4) c, (3, 4) d, (2, 5)
3. (5, 2)
4. a, Triangle
   b, Square
   c, Rectangle

9.10.
1. $99°$
2. 8th

10. (-2, 5)

3. 3 miles
4. $-5°$
5. -$ 38
6. $13°$
7. $8°$
8. $36°$ or $-106°$
9. $60°$
10. 140 or – 140
11. 10ft.
12. $16°$
13. Yes

342

14. 2260

15. Team A

16. $ 3766

9.11. Test

1. c
2. d
3. b
4. d
5. a
6. b
7. b
8. a
9. c
10. b
11. b
12. a
13. c
14. c
15. d
16. b
17. d
18. b
19. c
20. c
21. a
22. c
23. d
24. c
25. c

10.1.

1. Terminating
2. Repeating
3. Repeating
4. Terminating
5. Repeating
6. Terminating
7. Repeating
8. Repeating
9. Terminating
10. Repeating
11. Terminating
12. Repeating
13. Repeating
14. Repeating
15. Terminating
16. Repeating
17. Terminating
18. Repeating
19. Terminating
20. Repeating

10.2.1

1. <
2. >
3. >
4. =
5. <
6. <
7. >
8. >
9. <
10. =
11. =
12. >
13. <
14. >
15. =
16. =
17. <
18. >
19. <
20. >

10.2.2

1. -0.1, 1/10, 2/5, 1/2, 2

2. -2/3, -6/10, -1/2

3. -7/4, -2/5, 1/10, 5

4. -6, -5.01, -5, 0, 2

5. -7/4, 13/12, -0.65, -0.4, 1/10,

6. 14/5, 25/6, 10/2, 30/4, 64/7

7. -3.75, -2.833, -1.80

8. -10.4, -10.25, -6.833, 0.5, 10.4

9. -3.20, -2.40, 15/6, 6

10. -24.2, -18/7, 200/56, 215/36

11. 15/6, 6/5, 12/5, 16/5, 10

12. -16/5, 101/5, 104/51, 33/16, 5

10.3.

1. 0.25
2. 2.5
3. 1.43
4. 0.03
5. 9
6. 16
7. 9
8. 2
9. 5
10. 3
11. 3
12. 4
13. 0
14. 82
15. 25
16. 1
17. 2
18. 6
19. 12
20. 0
21. 6
22. 4
23. ¾
24. 0.645
25. 2.25
26. 4.5
27. 15
28. 2
29. 6
30. 36
31. 11/4
32. 3.333
33. Π
34. 14
35. 26
36. 42
37. 1.7778
38. 1.111
39. 1.111
40. 3.777

10.4.

1. 5
2. 1
3. 9
4. 4
5. 42
6. 100
7. 12
8. 16.2
9. 19
10. 12
11. 4
12. 5
13. 8
14. 2
15. 8
16. 8
17. 5
18. 6
19. 5
20. 1

10.5.

1. Triangle
2. Quadrilateral
3. Parallelogram
4. Square
5. Trapezoid
6. 12
7. 16
8. 12
9. 12
10. 20
11. 24
12. 15
13. 9
14. 9
15. 28
16. 8
17. 16
18. 6
19. 9
20. 22

10.6.

1. 8.6

2. 1
3. 14
4. -10.25
5. 11.1
6. 36
7. 16
8. -21.2
9. 10.5
10. -11
11. 4
12. -18
13. 4
14. -16
15. 4
16. 3
17. -10
18. 55
19. 4.28
20. -0.08
21. 4.09
22. -11.02
23. 0.6
24. 1.011
25. 0
26. 9

10.7.1

1. 30
2. 16

3. -18
4. 14
5. 72
6. 60
7. -72
8. 330
9. 165
10. 0
11. -2,460
12. 1,472
13. 819
14. -792
15. 4,608
16. -46,200
17. 10
18. -22
19. 240
20. -240
21. -44,800
22. -192
23. -1,331
24. 1

10.7.2

1. 5
2. 3
3. -3
4. -2
5. 30

6. 30
7. 30
8. -20
9. 46/125
10. 27
11. -5
12. -630
13. -5
14. 1
15. 9
16. -1
17. 1
18. -1
19. -100
20. 0

10.8.

1. 1.46
2. $4.19
3. 90.5
4. 98.9
5. 6.42 miles
6. 36
7. -3
8. 9
9. 7, -7
10. 14.04 in.

10.9. (Test)

1. d

2. b
3. a
4. b
5. c
6. d
7. a
8. b
9. b
10. a
11. a
12. a
13. b
14. d
15. c
16. a
17. b
18. c
19. d
20. c
21. d
22. c
23. a
24. b
25. d
26. c
27. c
28. b
29. c
30. a

Practice Exam 1

1. C
2. D
3. C
4. B
5. B
6. D
7. B
8. C
9. B
10. B
11. A
12. B
13. A
14. D
15. C
16. A
17. D
18. B
19. A
20. A
21. B
22. C
23. C
24. D
25. C
26. A
27. A
28. C
29. C
30. A
31. B
32. D
33. A
34. C
35. B
36. D
37. A
38. D
39. C
40. A

Practice Exam 2

1. C
2. C
3. D
4. A
5. B
6. D
7. C
8. B
9. D
10. A
11. B
12. B
13. C

14. A	1. D	29. C
15. B	2. A	30. B
16. A	3. C	31. D
17. B	4. B	32. C
18. A	5. D	33. D
19. D	6. B	34. B
20. A	7. D	35. A
21. A	8. A	36. B
22. C	9. C	37. C
23. B	10. B	38. B
24. A	11. B	39. C
25. C	12. B	40. A
26. B	13. A	Practice 4
27. A	14. C	1. a
28. C	15. D	2. c
29. B	16. B	3. b
30. D	17. A	4. d
31. C	18. B	5. a
32. A	19. A	6. b
33. C	20. C	7. d
34. D	21. D	8. b
35. B	22. B	9. a
36. A	23. D	10. c
37. B	24. B	11. a
38. D	25. C	12. c
39. B	26. A	13. d
40. D	27. D	14. b
PRACTICE 3	28. A	15. c

16. c
17. a
18. b
19. a
20. b
21. d
22. d
23. c
24. c
25. c
26. a
27. b
28. d
29. c
30. a
31. b
32. d
33. c
34. d
35. b
36. a
37. b

38. c
39. d
40. c

Practice 5

1. c
2. d
3. a
4. c
5. b
6. a
7. a
8. d
9. c
10. b
11. c
12. d
13. a
14. b
15. a
16. c
17. d
18. a

19. b
20. c
21. d
22. c
23. b
24. b
25. a
26. d
27. c
28. b
29. c
30. a
31. c
32. 150
33. 55
34. 55
35. 145
36. 190 – 200
37. 160 – 170
38. 55
39. b
40. c

www.ingramcontent.com/pod-product-compliance
Lightning Source LLC
Chambersburg PA
CBHW080450220526
45465CB00006B/2221